Heritage Conservation in Postcolonial India

Heritage Conservation in Postcolonial India seeks to position the conservation profession within historical, theoretical, and methodological frames to demonstrate how the field has evolved in the postcolonial decades and follow its various trajectories in research, education, advocacy, and practice.

Split into four sections, this book covers important themes of institutional and programmatic developments in the field of conservation; critical and contemporary challenges facing the profession; emerging trends in practice that seek to address contemporary challenges; and sustainable solutions to conservation issues.

The cases featured within the book elucidate the evolution of the heritage conservation profession, clarifying the role of key players at the central, state, and local level, and considering intangible, minority, colonial, modern, and vernacular heritages among others.

This book also showcases unique strands of conservation practice in the postcolonial decades to demonstrate the range, scope, and multiple avenues of development in the last seven decades. An ideal read for those interested in architecture, planning, historic preservation, urban studies, and South Asian studies.

Manish Chalana is an Associate Professor in the Department of Urban Design and Planning at the University of Washington with adjunct appointments in the Architecture and Landscape Architecture departments. He also serves on the faculty of South Asia Studies program in the Jackson School of International Studies. Additionally, Dr Chalana served as the director of the Graduate Certificate in Historic Preservation and co-directs the Center for Preservation and Adaptive Reuse. His work focuses on historic preservation planning, planning history, and international planning and development, particularly in his native India, primarily through the lenses of social justice and equity.

Ashima Krishna is Associate Director at the Purdue Policy Research Institute and Assistant Professor of Practice in the School of Interdisciplinary Studies at Purdue University. She is an architect and historic preservation planner whose research spans the management of historic urban landscapes and adaptive reuse of historic religious structures and landscapes, with a particular focus on intersection with community development issues and resulting policy challenges. Dr Krishna has examined issues related to historic preservation planning and urban conservation in the United States and India and continues to highlight the ways in which the historic built environment can be preserved, managed, and planned for.

Routledge Research in Architectural Conservation and Historic Preservation
Series Editor: Wen-Shao Chang

The Routledge Research in Architectural Conservation and Historic Preservation series provides the reader with the latest scholarship in the field of building conservation. The series publishes research from across the globe and covers areas as diverse as restoration techniques, preservation theory, technology, digital reconstruction, structures, materials, details, case studies, and much more. By making these studies available to the worldwide academic community, the series aims to promote quality architectural preservation research.

For more information about this series, please visit: https://www.routledge.com/architecture/series/RRACHP

Heritage Conservation in Postcolonial India

Approaches and Challenges

Edited by Manish Chalana and Ashima Krishna

Routledge
Taylor & Francis Group

LONDON AND NEW YORK

First published 2021
by Routledge
2 Park Square, Milton Park, Abingdon, Oxon OX14 4RN

and by Routledge
52 Vanderbilt Avenue, New York, NY 10017

Routledge is an imprint of the Taylor & Francis Group, an informa business

© 2021 selection and editorial matter, Manish Chalana and Ashima Krishna; individual chapters, the contributors

British Library Cataloguing-in-Publication Data
A catalogue record for this book is available from the British Library

Library of Congress Cataloging-in-Publication Data
Names: Chalana, Manish, editor. | Krishna, Ashima, editor.
Title: Heritage conservation in postcolonial India: approaches and challenges/edited by Manish Chalana and Ashima Krishna.
Description: Abingdon, Oxon; New York: Routledge, 2021. | Includes bibliographical references and index.
Identifiers: LCCN 2020033651 (print) | LCCN 2020033652 (ebook) | ISBN 9780367619947 (hardback) | ISBN 9781003109426 (ebook)
Subjects: LCSH: Historic preservation–India. | Cultural property–Protection–India.
Classification: LCC DS419 .H47 2021 (print) | LCC DS419 (ebook) | DDC 363.6/90954–dc23
LC record available at https://lccn.loc.gov/2020033651
LC ebook record available at https://lccn.loc.gov/2020033652

ISBN: 978-0-367-61994-7 (hbk)
ISBN: 978-0-367-62430-9 (pbk)
ISBN: 978-1-003-10942-6 (ebk)

Typeset in Sabon
by Deanta Global Publishing Services, Chennai, India

Contents

Contributors

Ananya Bhattacharya is a social entrepreneur with experience in empowering traditional artists and crafts persons for developing cultural enterprise. Ananya is an electrical engineer from Jadavpur University, Kolkata and a Commonwealth Scholar with a Master's in Sustainable Development from Staffordshire University, UK. She also serves as a director at banglanatak dot com – a social enterprise headquartered at Kolkata and working across India for pro-poor growth using culture-based approaches. Ananya is a member of the ICOMOS International Scientific Committees on Cultural Tourism and Intangible Cultural Heritage. She was a member of the Steering Committee of the Forum of accredited NGOs to the UNESCO 2003 Convention.

Kamalika Bose is an urban conservationist, researcher and educator based in Mumbai, and the founder of Heritage Synergies India. She is a Fulbright Scholar, an SAH-Getty International Fellow, and was formerly Assistant Professor at CEPT University, Ahmedabad. She obtained a Master's in Historic Preservation Planning from Cornell University. Kamalika has gained international work experience in areas of neighbourhood preservation and cultural heritage through positions at Historic Districts Council, New York, and the Cooper Hewitt Smithsonian Design Museum, New York. She has authored five books including *The Hoysala Legacy* (2019) and *People Called Kolkata* (2019) and her work attempts to bridge academia and practice in the domain of heritage conservation. Kamalika was the principal preservation consultant for The Cha Project.

Manish Chalana is an associate professor in the Department of Urban Design and Planning at the University of Washington with adjunct appointments in the Architecture and Landscapes Architecture departments. He also serves on the faculty of the South Asia Studies program in the Jackson School of International Studies. Dr Chalana also serves as the director of the Graduate Certificate in Historic Preservation and co-directs the Center for Preservation and Adaptive Reuse. His work focuses on historic preservation planning, planning history, and international planning and development, particularly in his native India, primarily through the lenses of social justice and equity. His recent projects have focused on urban transformations in Indian cities, and their impacts on historic neighborhoods and low-income communities. He also works on topics of resilience and disaster planning for at risk communities including in the Himalayas. His work has appeared in journals ranging from the Journal of Architectural Education, Journal of Heritage Stewardship, Planning Perspectives, Journal of Planning History and Journal of American Planning Association. He has also co-edited a volume titled *Messy Urbanism: Understanding the 'Other' Cities of Asia*, Hong Kong University Press, 2016.

Patricia Tusa Fels is the founder and principal architect of PTF Architects. For over thirty years she has been involved in conservation projects in the United States, Italy, and Asia. She is a graduate of the University of California, Berkeley School of Architecture. Her interest in Islam and mosques began during her years at Berkeley where she first studied Islamic architecture. In complement to her architecture practice, Patricia writes extensively about cities, change, and the importance of place, especially the role of vernacular buildings. Her writing has been featured in journals, magazines, newspapers, on the web, and in books. Her book *Mosques of Cochin* (Mapin, 2011) is the result of five years of research in Kerala. Her second book, *Monsoon Mosques* (Mapin, 2020) expands the exploration of timber mosques to Indonesia and Malaysia.

Divay Gupta is a leading conservation architect in India and heads the Architectural Heritage Division of INTACH, New Delhi. An alumni of ICCROM, University of Birmingham School of Planning and Architecture, he has been striving to help better manage and conserve the cultural resources in his country for more than 20 years. He has had the privilege and opportunity to be part of a number of prestigious projects in the UK, USA, India, Afghanistan, Nepal, and Cambodia and has participated in joint UNESCO and ICOMOS missions to world heritage sites as a conservation expert. His restoration projects in Ladakh have won the SA UNESCO awards of Merit and Excellence. He has been invited for special lectures and juries at various national and international conferences, seminars and colleges of Architecture across India. He is a visiting faculty and a member on the board of studies of Department of Architecture Conservation at SPA New Delhi, and a faculty member at the School of Architecture at Delhi Technical Campus, G. Noida.

Priya Jain is an assistant professor of architecture and associate director of the Center for Heritage Conservation at Texas A&M University. An architect licensed both in India and the US, she previously held leadership positions at two Boston-area preservation firms. Her portfolio includes work on a number of historically significant buildings including Trinity Church in Boston and the Jewett Arts Center at Wellesley College. Her current research and teaching lie at the boundaries of design, preservation theory and practice in a transnational context. She is particularly interested in preservation of institutional sites and buildings of the recent past. Her work has been shared at various conferences and published in the *Journal of Architectural Education*, among others. She serves as Field Editor (Architecture) for the Getty Conservation Institute's Abstracts of International Conservation Literature and is on the Board of the Association for Preservation Technology (APTI) Texas Chapter and South Asia Chapter.

Shikha Jain has vast experience in cultural heritage that covers conservation, world heritage and museum planning with over 50 projects across India and overseas, largely realised through her organisation DRONAH. She has led several conservation projects funded by the Getty Foundation and the World Monuments Fund. Dr. Jain has advised government organisations in Singapore, Malaysia, UAE and Myanmar, and UNESCO Offices at Jakarta, Indonesia and Myanmar on world heritage nominations. She has also worked as a consultant for UNESCO, New Delhi and represented India as Cultural Heritage Expert on the UNESCO World Heritage Committee. Her urban conservation projects for Jaipur have received two awards from the Housing and Urban Development Corporation and recognised as a model of best practice by the Ministry of Urban Affairs, India.

Swapna Kothari is a practicing conservation architect and preservation professional. An advocate for the built environment and cultural traditions, she conducts documentation works for historic buildings and studies how institutions aid each other with conservation tools and theory. Apart from conserving built spaces, she is an educator of young minds and a key member in reviving a non-profit that looks into different avenues of heritage conservation. Her professional and academic commitments have taken her from palaces in India, to temples in Cambodia, and mansions in New York State.

Ashima Krishna is Associate Director at the Purdue Policy Research Institute and Assistant Professor of Practice in the School of Interdisciplinary Studies at Purdue University. She is an architect and historic preservation planner whose research spans the management of historic urban landscapes and adaptive reuse of historic religious structures and landscapes, with a particular focus on intersection with community development issues and resulting policy challenges. Dr Krishna has examined issues related to historic preservation planning and urban conservation in the United States and India and continues to highlight the ways in which the historic built environment can be preserved, managed, and planned for.

Churnjeet Mahn is Reader in English Literature at the University of Strathclyde and a fellow of the Young Academy of Scotland (Royal Society of Edinburgh). Her research investigates the history and practice of travel with special reference to race and nationalisms. Her early work investigated the competing discourses of Orientalism and Hellenism in Greece in the wake of its independence from the Ottoman Empire which was published as *Journeys in the Palimpsest: British Women's Travel in Greece 1840–1914* (Routledge). Her work in Punjab investigated how partition marked the end of Islamic influence in Northern India. In both contexts, she has been interested in how Islamaphobia has been used to create a national 'other'. She recently completed a large Arts and Humanities Research Council project entitled Creative Interruptions and she is currently running a British Academy grant entitled Cross-Border Queers: The Story of South Asian Migrants to the UK.

Yaaminey Mubayi is a social and institutional development specialist with approximately twenty years of experience working with a wide spectrum of organisations ranging from international multilateral agencies (UNESCO, the World Bank, ADB), private sector CSR initiatives (the Nabha Foundation) and state and central government agencies (Ministry of Culture, Government of India, state governments of Punjab, Orissa and Uttarakhand). She has a Master's in social policy (NGO Management) from the London School of Economics and a doctorate in history from the Jawaharlal Nehru University, New Delhi. She has received the junior and Senior Research Fellowships from the University Grants Commission, Government of India, a DAAD Fellowship for doctoral research at the South Asia Institute, University of Heidelberg, Germany as well as an archival grant from the Charles Wallace Trust, UK. More recently she has received the Senior Academic Fellowship from the Indian Council for Historical Research (ICHR), Government of India. She is currently a visiting faculty at the Department of Conservation Studies, School of Planning and Architecture, New Delhi.

Gurmeet S Rai is an architect with specialization in heritage conservation and management. She is among the first-generation conservation architects in India and has worked in the area of architectural conservation, management plans for world heritage sites, urban conservation strategies for historic settlements, sustainable cultural heritage tourism and preparation of advisory and policy documents. She prepared the Site Management Plans for the World Heritage Sites of Red Fort in Delhi, Ellora Caves in Maharashtra, the conservation management plan for Gobindgarh Fort in Amritsar, Grand Trunk Road in Punjab and conservation plans for historic buildings in the Golden temple in Amritsar. She has received the Award of Distinction from the UNESCO Asia Pacific Architectural Heritage Awards in the years 2002 and 2004, following which she has been on the jury. She is currently a member of the Steering Committee of TERRA 2021, World Congress on Earthen Architectural Heritage (Getty Conservation Institute) and serves on several national advisory and academic bodies related to heritage.

Saptarshi Sanyal, a doctoral candidate at University College London's Bartlett School of Architecture, is on research leave from his full-time academic appointment at New Delhi's School of Planning and Architecture, where he has been assistant professor since 2013. Formerly in the ASI (2010–13) he was involved in heritage conservation and management in over fifteen states in India. Since 2019, he has been a tutor in Bartlett's BSc architecture program. An educator in architecture and design, its histories and visual storytelling, Saptarshi has contributed through tutorials and workshops in over twenty institutes all over India. He is an independent photographer with short- and long-term bodies of work exhibited and published in three continents (c. 2010–16). Saptarshi has written extensively on the built environment's knowledge systems in heritage and architectural history. His PhD is funded by the UK government's Commonwealth Scholarship Commission (2018–21).

Priyaleen Singh is a professor in the Department of Architectural Conservation at the School of Planning and Architecture, New Delhi. She was awarded the Charles Wallace India Trust scholarship to do her MA in Conservation from the Institute of Advanced Architectural Studies, University of York, U.K. and subsequently got the Commonwealth Scholarship to do her D.Phil from the same institute in 1998 on 'Changing Attitudes to Design with Nature in the Urban Indian context'. She has been a summer fellow at the Dumbarton Oaks Institute, Washington D.C. As a practicing conservation architect and a landscape architect, she has worked on several urban conservation and historic landscape conservation projects and continues to be concerned with contemporary landscape design and urban conservation issues in India.

Amita Sinha is a former professor in the Department of Landscape Architecture at the University of Illinois at Urbana Champaign and has taught in the Department of Architecture and Regional Planning, IIT Kharagpur and in the Humanities and Social Sciences Department at IIT Gandhinagar in India. She is the author of *Landscapes in India: Forms and Meanings* (University Press of Colorado, 2006; reprinted by Asia Educational Services, 2011), *Cultural Landscapes of India: Imagined, Enacted, and Reclaimed* (University of Pittsburgh Press, 2020), editor of *Landscape Perception* (Academic Press, 1995), and co-editor of *Cultural Landscapes of South Asia: Studies in Heritage Conservation and Management* (Routledge, 2017). She was a senior Fulbright researcher at the Indian National Trust for Art and Cultural Heritage (INTACH) in New Delhi in 2009 and is the recipient of Fulbright-Nehru Academic and Professional Excellence Award Fellowship in 2018–19. She received the National Merit Award, American Society of Landscape Architects for the Cultural Heritage Project on Taj Mahal, India, in 2001 and Environmental Design Research Association (EDRA) Achievement Award in 2018.

Aishwarya Tipnis is an architect, educator, and conservation professional based in New Delhi India. An alumnus of the School of Planning and Architecture, New Delhi and the University of Dundee, Scotland, she has nearly two decades of experience in the field of architectural and urban conservation and has pioneered several urban conservation and building restoration projects in India. Aishwarya is the recipient of the highest civilian honour in the culture sector, the insignia of *Chevalier des Arts et des Lettres* (Knight of the Arts and Letters), by the Government of France in 2018 for her outstanding commitment to the preservation of French Heritage in West Bengal. Recognized as a Global Cultural Leader by the European Union in 2016, she is also the recipient of the UNESCO Award for Heritage Conservation in the Asia-Pacific Region (Award of Merit: Mahidpur Fort & Honourable Mention for Doon School, Dehradun) in 2016.

Michael A. Tomlan directs the Graduate Program in Historic Preservation Planning in the College of Architecture, Art and Planning at Cornell University. Tomlan served as chair of the Senior Board of Advisers to the Global Heritage Fund (Palo Alto, California), reviewing nominations for and evaluating conservation projects in India and other countries in Asia. He consulted on projects for the World Monuments Fund, the J. Paul Getty Trust, is currently chair of the board of Yosothor (Cambodia), and serves as a project director for the National Council for Preservation Education. He is the author of *Historic Preservation: Caring for Our Expanding Legacy* (Springer, 2014), and the editor of *Preservation of What, For Whom?* (NCPE, 1997).

Sakriti Vishwakarma is a conservation architect from India, now based in the USA. Her passion for heritage preservation led her to pursue a M.Arch in architectural conservation from SPA, Delhi-India after her undergraduate degree in architecture. She practiced and taught in India before she completed her Master's in real estate from University of Washington in 2020.

James L. Wescoat, Jr. is Aga Khan Professor Emeritus of Landscape Architecture and Geography at the Massachusetts Institute of Technology. He also co-directed the Norman B. Leventhal Center for Advanced Urbanism. He earned a Bachelor of Landscape Architecture degree from Louisiana State University and graduate degrees in geography from the University of Chicago with an emphasis on water resources. His research concentrates on water systems in South Asia from the site to river basin scales. At the site scale, Professor Wescoat has focused on historical waterworks of Mughal gardens in the cities of Agra, Delhi, Lahore, and Nagaur. At the regional scale, he has led projects in Gujarat, Maharashtra, and Punjab.

Foreword

A series of paradoxes underlie conservation practice, some laced with irony. For example, we work diligently advocating for historic sites so that people will see them as more than functional objects, but rather as expressions of a culture, imbued with values and meanings that warrant their preservation: beauty, the commemoration of important people and events, connecting people to their heritage, telling an important story about how the community got there. And then when we succeed, when people see and absorb the significance a site has for themselves and others, we all too often get to watch as those meanings become a justification for the site's destruction. We live in a world of multiple, conflicting narratives and cultural and community identities and goals, as we have seen in recent years with the destruction of the Bamiyan Buddhas in Afghanistan, or the Mostar Bridge in Bosnia and Herzegovina. It takes a certain acceptance of complexity, and a deliberate collaborative process, to achieve the rarer example of the site whose meanings and narratives have both changed and expanded. Southern plantations in the US are one example of this being done; central to a brutal economic system dependent on slavery, preserved in many cases in the first place because of their architectural splendour and connections to important white families, but now interpreted by descendants of both the residents of the 'big house' and the enslaved.

A second paradox has to do with the change of use that often accompanies the act of preservation, especially for sites of heritage tourism, in which the economic benefits to a community become both welcome benefit and trap. An enormous global industry, heritage tourism is first about money, second about tourism, third about status, and lastly about heritage (Boniface and Fowler 1993). It is about the provision of a consistent, pleasurable experience to people who are often not culturally connected to the site; about maximizing the economic return to a community through jobs, ticket sales, and ancillary purchases of souvenirs, lodging, food, and drink. It is even about demonstrating the importance of a community's contributions and heritage on the global stage. All too often, only when these other things are satisfied are the needs of the site itself, and its innate associations with the community in which it sits, considered. Admittedly, many communities would not have the resources to preserve if tourism did not provide the funds, and many people who otherwise would not care about the sites become invested because of the flow of economic benefits. However, as sites transition away from the roles they have played over decades or centuries, there is too often a loss of something essential, whether it is called character, or integrity, or authenticity. This too is a complicated question, as we see playing out dramatically in 2020 with the decision by the Turkish Supreme Court that the conversion of Hagia Sofia in Istanbul from mosque to museum in 1934 was invalid; and President Erdogan's immediate announcement that the site would re-open for prayers as a mosque within weeks. A building constructed as a church and converted to a mosque some 900 years later – which it was for 500 years – then for 90 years became a museum and a symbol to many of inter-cultural heritage and cooperation. International organizations are now suggesting that Hagia Sofia's historic designation is at risk because it might return to its original use as a house of worship.

A third paradox, akin to the first, has to do with the association of heritage and power. Even beyond identity, a certain heritage is seen as conferring legitimacy on the ruling party. Mexico is rife with symbols of its pre-Columbian heritage, not just in preserved sites, but in images used by the government, along with large corporations and others, to claim their positions in a millennia-long lineage (Canclini 1995). The irony being, of course, that the government and the ruling class in Mexico have been composed mostly of descendants of European (post-Columbus) settlers, while the descendants of the pre-Colombian peoples remain largely economically and politically disadvantaged. This example underscores the degree to which preservation, like most design in the public realm, is political—its outcomes dependent on the community will, on government regulations and incentives, on consensus around narratives and identities. These are difficult enough to negotiate when a cultural and ethnically homogeneous community is debating its heritage. It is even more fraught when multiple communities co-exist in the same area, 'reading' sites in very different ways.

In India, the well-known example of the Babri Masjid in Ayodhyā continues to fester: in 1992, a mob destroyed a mosque (purportedly dating from the 16th century) because it was allegedly constructed not just on the site of a Hindu temple that had been demolished in the process, but on the birthplace of the Hindu god, Rama. What to do with the site triggered legal cases that found their way to the Indian Supreme Court, which in 2019 ruled that the site was to be given to a trust for the construction of Hindu temple. There is no question that the pro-Muslim biases of the British, and anti-Muslim biases in the present national and state governments in India, have played determinant roles in the saga.

India's population is the second largest on earth, and its cultural diversity is demonstrated by the fact that its inhabitants speak, by some counts, 780 distinct languages and dialects, 30 by over a million people each. Taking a train across India, a profound experience by any measure, reveals some of this diversity, as the place names in each station not only change language, but often even their alphabet. India has given birth to Buddhism, Hinduism, Jainism, and Sikhism, and still contains the third-largest Muslim population of any country and almost 30 million Christians. And while vanishingly small today, the first Jews arrived in India as early as the 6th century BCE, with a significant influx after the fall of the second temple, establishing communities with still-extant synagogues in at least three regions of the country.

The lands that are part of India today have had multiple histories of migration and conquest for at least 4,000 years. The Dravidian culture arose sometime after 5,000 BCE when agricultural settlers from the Middle East mixed with local aboriginal peoples, resulting in the linguistic and ethnic communities that still dominate South India. Perhaps a millennia later, peoples from what is now Eastern Europe and the steppes of Central Asia moved into the northern parts of the subcontinent, most notably the Indus Valley, and gave rise to the Indo-Europeans, including the Aryans.

Over the centuries, various invaders, from Alexander the Great to the Mongols, as well as native rulers such as Ashoka, controlled territories and established vast empires in present-day India – perhaps most dramatically in the cases of the Mauryan Empire of 322 to 185 BCE, and the Mughal empire of 1526 to 1707, both of which grew to include most of present-day South Asia: large stretches of India, Pakistan, Afghanistan, and Bangladesh. From the 15th century on, following the rapid rise of the colonial ambitions of Europe, parts of India were also ruled by the Dutch, Danes, Portuguese, French, and of course, British. The nation took its present form only with partition following independence in 1947, the abdication of the last of over 580 putatively independent princely states in 1949, and evacuation of the last European colony, Goa, in 1961.

During the Cold War, India was a founder of the non-aligned movement – in partnership with Yugoslavia, Indonesia, and Egypt – which allowed India to interact selectively with both sides while it became the world's largest democracy. The principles of Nehruvian governance, which included a secular state that fought against the deeply ingrained caste system and attempted to bring all religious and ethnic factions and castes into the political system, dominated the political scene for decades but has been in stark retreat the past few years with the rise of the BJP and the Hindutva movement. The monuments of that period, indeed much of the 'architecture of independence,' as one might call the extraordinary output of architecture and planning in the 1950s through 1970s, are now at risk from two directions. First, as Ramachandra Guha has noted, Indians tend to see everything since 1947 as current events. Legitimizing preservation is challenging if a site's period of significance is not seen as historically important to the development of the nation – and indeed many important modernist sites, such as the Pragati Maidan exhibition grounds or the Chanakyapuri Theatre, have been demolished, while others, such as the India International Centre, were radically altered. Secondly, the social and political aspirations behind many of the works, and of both modernism and the ruling Congress Party, are now viewed as problematic at best.

The purpose of this overly simplified history is to suggest the extraordinary context for preservation in India. The sites that densely dot the landscape come from different civilizations, most of which are known, but many of which remain unknown. Their significance, the narratives constructed around them, are dramatically different depending on which community is telling the story. The principles and practices of preservation in India are largely, still, taken from either the colonial powers, especially the British, or from international debates held elsewhere – although voices that seek to present a more 'Indian' approach are increasingly being heard.

All the paradoxes described earlier can be found at play in India today. And like many places with a long history, the debates are fuelled by enmities, or at least conflicting understandings of that history, which may go back to events from generations past, and yet are as current as their last retelling. Like much of the world in 2020, culture wars are front and centre in political debates in India, too.

For several decades after the founding of formal university programs in historic preservation in the US and elsewhere in the 1970s, the major texts in the field had a Western focus, just as the international preservation conventions came from cities such as Athens or Venice. Gradually, the lens expanded to give us books that sampled conservation in different countries, although still often through the lens of a single western-based NGO, such as the World Monuments Fund, or from the perspective of international governmental organizations based in the west, such as UNESCO or ICOMOS. Only recently has it been possible to produce a book like the one here, in which multiple voices describe how a single country located in the Global South, outside the power centres of the discipline, is wrestling with and putting together its preservation history, its ideas about what preservation can and should be, its path. While the essays address difficult issues and circumstances, and frequently present cautionary examples, their cumulative voice speaks to maturity and understanding that is eminently hopeful for the extraordinary cultural resources of India.

Jeffrey M. Chusid
Associate Professor and Chair, Department of City
and Regional Planning, Cornell University

References

Boniface, Pricilla, and P.J., Fowler. 1993. *Heritage and Tourism in the Global Village*. Routledge.

Canclini, Nestor Garcia. 1995. *Hybrid Cultures: Strategies for Entering and Leaving Modernity*. University of Minnesota Press.

Acknowledgments

A work of this magnitude is a reflection of efforts and assistance by numerous individuals and organizations, without whom this work would have remained incomplete. First off, we are thankful to all of the contributing authors in the volume who have so generously shared their expertise through their work. Understanding heritage conservation in the past seven decades in post-colonial India is a daunting endeavour to which no single volume can fully do justice. Collectively, the remarkable group of scholars and practitioners contributing to this work have highlighted some avenues of diversification for the field, from being largely monument-focused to a broader appreciation of vernacular, modern, and intangible heritage. Thanks to their insights, we believe this work will contribute to the learning of academics, students, and practitioners in heritage conservation in India and those interested in these topics.

We are also incredibly grateful to Routledge UK for recognizing the merit of our project and shepherding it through the review and publication stages. We are especially grateful to Grace Harrison, former Editor, Landscape Architecture, Built Environment Research & Product Design, and Julia Pollacco, former Editorial Assistant, for their guidance throughout the process. This project would not have reached its finishing line without the invaluable assistance of Lauren Nickodemus, Edward Gibbons, and Christine Bondira. To them we remain appreciative.

We are grateful to the University of Washington for awarding a Royalty Research Fellowship to Manish Chalana to support the initial research that led to the production of this volume. The American Institute of India Studies (AIIS) generously supported his broader scholarship on historic preservation in India through a Senior Long Term Research Fellowship, which informed this work in countless ways. The UW's College of Built Environments provided funds from their Johnston-Hastings Publication Endowment, which made the final production of the volume possible. Thanks are also due to the South Asia Center in the Jackson School of International Studies at the University of Washington for their overarching support and intellectual community.

Our careers in historic preservation in India and the US have been shaped by many individuals who have refined our thinking on the topic. We remain indebted to A.G.K. Menon, Ann Komara, Cecilia Rusnak, Chuck Wolfe, Dan Nadenicek, Gita Diwan Verma, Jeffrey Chusid, Kathryn Gleason, K.T. Ravindran, Mary Woods, Michael Holleran, Michael Tomlan, Nalini Thakur, Rajat Ray, Sohyun Park Lee, and Tom Yahner. Interactions with them over the decades have provided opportunities for feedback and reflection that have enriched our work. Manish would like to thank the colleagues, students, and friends at the University of Washington who supported this effort in multiple ways, especially Amber Piona, Anne Vernez Moudon, Bob Freitag, Bob Mugerauer, Dan Abramson, Radhika Govindrajan, and Sunila Kale. He reserves a special thanks for Steven Goodreau and William Stanford for their unwavering support from the start to finish of this work. Ashima would like to thank colleagues, students, and friends at the University at Buffalo for providing guidance throughout the publication and editorial process, including Daniel Baldwin Hess, Ernest Sternberg, and especially Samina Raja for her invaluable advice and support for the project. Ashima is especially thankful to Krishna Jayant and Neeraja Krishna, without whose unwavering support and shared childcare duties, this project would not have been the same, or on time! We remain deeply indebted to all who supported this project in big ways and small. Thank you.

Manish Chalana and Ashima Krishna

Introduction: Untangling heritage conservation in postcolonial India

From colonial to global times

Manish Chalana and Ashima Krishna

Introduction

India is an ancient civilization whose cultural heritage ranges over millennia and exhibits a richness and diversity that is nearly unparalleled. However, the formal profession of heritage conservation (or historic preservation as it is known elsewhere), is a relatively recent phenomenon, established under the British Raj in the second half of the nineteenth century with the creation of the Archaeological Survey of India (ASI). Since that time, and particularly in the period following India's independence in 1947, the field has made significant theoretical and methodological advances and opened its doors to the involvement of both private sector and non-governmental organizations. Despite this maturation, the field in many ways remains marginal in practice and not successfully integrated with planning and policy. India's rapid urbanization and development in recent decades, fuelled in part by globalization and facilitated by the central government's urban programmes, threaten historic sites, the bulk of which remains unprotected. Several other major challenges confront the field today, ranging from a rise in communalism and Hindu nationalism to natural hazards, terrorism, and climate change.

This book positions heritage conservation within a range of theoretical and practical frames to demonstrate how it has evolved along various trajectories in the postcolonial decades. The main objective of this work is to present a rich and compelling range of projects and perspectives from scholars and professionals that showcase the evolution of the field and fill in the existing gaps in scholarship on those topics. In this introductory chapter, we provide an overview of the historical development of the formal profession of heritage conservation in India, from the creation of the Archaeological Survey of India to present times. We then introduce existing literature on the topic of heritage conservation in India as a context for the works discussed in this volume. Next, we identify four major themes that emerge from the chapters that broadly encompass institutional frameworks, sustainable approaches to practice, continuing challenges, and policy implications for heritage conservation in India. It is our aim that this volume will spur a new round of scholarship providing further critical assessment of the field of heritage conservation in the Indian context.

The Archaeological Survey of India: a thumbnail history

The formal practice of heritage conservation was initiated in 1861 by the creation of the Archaeological Survey of India (ASI) by the British colonial government. Its predecessor, the Asiatic Society of Bengal (current name: Asiatic Society), founded nearly a century earlier, in 1784, primarily focused on the textual sources of India's history and culture. Nevertheless, the society was instrumental in showcasing the depth and richness of Indian history, and its prominence paved the way for the British colonial interest in Indian archaeology and heritage conservation. Even as the Asiatic Society of Bengal did not initiate any archaeological excavation, their exegetic methodology would shape the ASI's approach well into the postcolonial era (Johnson-Roehr 2008).

The ASI's mandate used a very broad definition of archaeology, encompassing multiple forms of custodianship of cultural heritage in the built environment. Alexander Cunningham, an engineer by training, served as its first Director General. The ASI took on the task of documenting and protecting India's antiquity by conducting geographical surveys of historic and archaeological sites. The agency's work was further legitimized in 1904 by the passage of the Ancient Monuments and Preservation Act, which provided protections and acquisition rights for both ancient monuments and other objects of archaeological, historical, or artistic interest. The Act also placed controls on the trafficking of antiquities and on unauthorized excavations. Notably, its jurisdiction was set to include properties in private ownership.

The agency produced a landmark handbook of best practices – the *Conservation Manual* – in 1923, under the guidance of then Director General, John Marshall (Marshall 1923). The document provided technical instructions on building conservation with the goal of ensuring that conservation practice was uniform across the empire. In that regard, the *Conservation Manual* was intended to be a "prescriptive colonial text, setting down stringent rules for the practice of monument preservation in a colony, and thus constituted a text of authority" (I. Sengupta 2013). There was some indigenous resistance to the rules of site management as laid out in the *Conservation Manual*, particularly for sacred sites that were still in use; for these, the codes had to reconcile local expectations with the ASI's preferred conservation approaches. But for the most part, the rules and regulations of the *Conservation Manual* were uniformly enforced, making it one of the most influential documents guiding practice in the colonial – and ultimately into the postcolonial era (J. Sharma 2012). The agency kept meticulous records; in addition to annual reports, it published reports on various excavations and on artefacts associated with specific excavations. These continue to be invaluable sources on topics of archaeology and heritage conservation throughout India.

After independence, the ASI continued to dominate the formal practice of heritage conservation, focusing largely on monuments and archaeological sites, using a scientific restoration methodology prescribed in the *Conservation Manual*. In 1951, the passage of the Ancient and Historical Monuments and Archaeological Sites and Remains (Declaration of National Importance) Act clarified the concept of "national importance" of monuments and sites. However, this Act, along with the colonial-era 1904 Act, was superseded in 1958 by the Ancient Monuments and Archaeological Sites and Remains Act (AMASR).[1] Although the basic mandate to preserve monuments, archaeological sites, and objects of national importance remained the same, the AMASR was passed in the interest of aligning the law better with the constitution of the newly independent India and assigning more powers to the Director General of Archaeology (A. Jain 2010). The law's 2010 amendment is significant in its attempt at maintaining the context of protected sites by creating prohibited and regulated zones around the protected sites of 100 and 200 metres respectively (Figure 0.1).

Today the ASI is housed under the Ministry of Culture, and continues to further its original mission of protecting the country's antiquities; currently, the agency protects 3,677 monuments and archaeological sites ranging from palaces and forts to archaeological mounds, with an additional 5,000-odd sites protected by the state governments. It carries out its work through 24 zonal offices (or circles), further divided into sub-circles with a superintending archaeologist as the appointed head of the circle. Historically, the central, state, and local agencies have had trouble coordinating on conservation issues across these hierarchies, which impacts the agency's effectiveness in carrying out its vision (Krishna 2017, 175). Additionally, given India's long history and diversity of cultural records, far greater numbers of historic sites remain unprotected compared to those under the ASI's jurisdiction. The agency has to make do with limited appropriation of funds from the central government funnelled through the Ministry of Culture, which receives only 0.2% of the national budget (compared to 20% for defence) for all of the entities under its jurisdictions, of which the ASI is but one. Partly due to this reason, the ASI cautiously expands its listing of protected sites, which thus largely remains limited to grand monuments and archaeological sites older than 100 years. As a result, the set of ASI sites do not fully capture the diversity of the country's resource types, time periods, geographical regions, and associations with cultural groups. This also leaves the bulk of non-high-style and more recent heritage underappreciated and outside the recognition and protections afforded by ASI designation.

Figure 0.1 Haveli Khajanchi, Dariba Kalan, Shahjahanabad, exhibiting state of disrepair and neglect. This once opulent residence is believed to be that of Shah Jahan's *khajanchi* (treasurer). Many such partially abandoned forlorn *havelis* dot the urban fabric of Shahjanabad. Source: Manish Chalana.

Beyond the ASI: INTACH, UNESCO, and the States

Thankfully, the field of heritage conservation in India has expanded beyond the ASI in recent decades, most notably with the creation of the Indian National Trust for Art and Cultural Heritage (INTACH) in 1984. Similar to the National Trust in England and the United States, INTACH is a parallel non-governmental preservation organization focusing on advocacy, education, and outreach on matters of heritage conservation. INTACH conceptualizes heritage in three categories: built, natural, and intangible (or "living heritage"). The organization is assisted in its efforts by approximately 200 chapters spread throughout the country. Despite its lack of jurisdictional powers, INTACH's impact in shaping the conservation profession in India is far-reaching, even as some of the theoretical and methodological strides they represent can take time to infiltrate into broader practice. Overall, though, INTACH is fundamental in the effort to present a more representative past that aligns better with India's historical development and cultural and socio-economic diversity.

Since the 1990s heritage conservation in India has also witnessed a greater role of global heritage agencies in shaping the practice, seen in the continual rise of World Heritage sites in the country. As of 2019, there are 37 World Heritage sites in India (29 cultural; 7 natural; and 1 mixed). While this number may seem high, it remains well below those of much smaller, less populous European countries such as Italy and Spain, with 54 and 47, respectively. UNESCO recognized and began addressing its Eurocentric bias in World Heritage listings with the 1994 creation of the *Global Strategy for a Balanced, Representative and Credible World Heritage List*[2]; indeed, the majority of World Heritage sites in India have been inscribed since the 1990s. However, much of UNESCO's work in India has until recently focused on the crown jewels of cultural resources, the likes of the Taj Mahal and Agra Fort, showcasing grand architecture and a glorious and distant past associated with the elites. Recent developments are promising, however, including the inscription of the walled city of Ahmedabad in 2017 as a World Heritage City, marking a recognition of non-high-style and lived-in environments.

The 1990s also saw a greater opening up of India to global commerce and capital, ushering in a new era of rapid urban transformation fuelled by several national urban projects related to infrastructure development and poverty alleviation. However, rapid urbanization has come at the cost of a loss of unprotected and undocumented historic sites (Figure 0.2), including traditional neighbourhoods, which have been transformed under pressures of both urbanization and tourism. Some cities have begun their own efforts to take stock of their vernacular heritage, using a more local lens. Ahmedabad and Mumbai have been leaders in this movement, each creating a municipal-level heritage regulation that protects the historic fabric in designated historic precincts. However, these are insufficient in recognizing a range of cultural resource types, particularly industrial and those related to the working classes, or ensuring that this preservation work does not actively displace those very communities they were built to serve (Chalana 2009).

Recent scholarship

The literature on heritage conservation in India has been growing steadily since India's independence, with a parallel expansion in practice and methods. The bulk of this literature can be broadly sorted into three categories. The first set focuses on colonial and postcolonial archaeology generated by the Archaeological Survey of India (ASI) and by practitioners working on ASI projects. Additionally, independent scholars who study the ASI have also contributed to this set. The second set, with a more limited circulation, is generated by INTACH and by practitioners reporting on their projects. Finally, the remaining literature is produced by academics, public scholars, and practitioners on a range of theoretical, methodological, and practical topics in the field. In this section, we provide a quick overview of a sampling

Figure 0.2 Khirki Masjid, an ASI-protected monument is now engulfed by a much
transformed Khirki Village. It went through some hasty restoration work
in preparation for the Commonwealth Games that Delhi hosted in 2010,
which the ASI eventually suspended as pressure to reassess the restoration
work by INTACH and other groups mounted. Source: Manish Chalana.

of publications within each set of literature. This is by no means comprehensive, as the literature in each category is extensive, and covering it fully
is beyond the scope of this work. Our goal is to familiarize readers with a
range of topics in each category and to situate our work within the existing
body of literature.

ASI publications

The Archaeological Survey of India's main output comprises books and
reports showcasing their sites and approaches, with a focus on excavations and cataloguing. The agency publishes annual reports that track the
work and expenses on projects, among other routine record keeping. These
reports were initiated in 1862 by Cunningham, the first Director General of
the ASI. Additionally, there are numerous books focusing on ASI excavations through the decades on a variety of sites. Examples provide a sense of
the specificity found in these books; one covers details on excavation work

carried out from 1985–91 on a large Buddhist complex in Lalitagiri, Odisha (Patnaik 2016), while another details excavations in Udayagiri, Andhra Pradesh (Bandyopadhyay 2007). Another reports on the 20 rock-cut caves from the 5th century CE near Vidisha, Madhya Pradesh (Bandyopadhyay 2004). Publications cataloguing artefacts from various ASI excavation sites include examples such as the thousands of terracotta figurines of humans and animals from archaeological mounds in Rajghat, north of Varanasi near Sarnath, which were accidentally discovered during colonial times (Mani 2012).

During James Burgess's tenure as the second ASI Director General (after Cunningham), the ASI began publishing an annual journal with the thoroughly Victorian name *The Indian Antiquary: A Journal of Oriental Research in Archaeology, History, Literature, Languages, Philosophy, Religion, Folklore, &c &c. The Indian Antiquary* started publication in 1872, with a break from 1933–1937, and was then republished until 1947 under the name *The New Indian Antiquary,* then revived one final time from 1964–1971. The journal was meant for sharing ideas and disseminating information on a broad range of topics (including conservation) between scholars and practitioners within India and abroad (Archaeological Survey of India 1872). A scan of journal issues reveals details on various historic and archaeological sites; for example, excavations in Udayagiri are discussed in the 1884 issue; inscriptions in Devanagari script on pillars in one cave from the 11th century are described in great detail (*The Indian Antiquary*, v.13, 1884).

Academic and public scholars have also researched the ASI, from its early years (Keay 2011), through to a more recent retrospective on the agency's 150 years of operation (G. Sengupta, Lambah, and Archaeological Survey of India 2012). Pant's detailed archival account of the efforts of the English East India Company and the British Crown is a more recent and comprehensive historical understanding of how heritage conservation was (or not) carried out in the period between 1784 and 1904. He describes efforts prior to the formation of the ASI, and then focuses on the formation and activities of the agency, specifically through the leadership of Alexander Cunningham, H.H. Cole, James Burgess, and John Marshall (Pant 2012). While presenting a rich tapestry of detailed historical information, this work falls short of any analysis or critique of the activities that took place in the pre-independence period.

Similarly, much of the prevalent literature remains uncritical of the ASI's effectiveness as the largest and most important stakeholder in historic preservation in India, with a major impact on practice. There are some notable counterexamples, however, especially more recently. These include Johnson-Roehr's investigation of how the legacies of colonial-era ASI continue to colour postcolonial practices, particularly in privileging Hindu holy sites and perpetuating violent disagreement between Hindus and Muslims, as with Babri Masjid in Ayodhyā (Johnson-Roehr 2008). Krishna's work highlights the historical and contemporary dissonance between local, state, and central agencies in administering and enforcing heritage conservation-related issues. Using the case study of Lucknow, Krishna critiques the local and state policy impacts of the ASI and highlights the inability of the agency to enforce their own laws locally without the cooperation and assistance of local administrations (Krishna 2017). Additionally, there exist unresolved tensions between the ASI's bureaucratic roles and expectations, and its cultural and scientific pursuit of archaeology. The agency's deep-rooted administrative bureaucracy does not always readily align with the scientific methods used in the ASI's routine archaeological practices at excavation sites (Chadha 2007). Recently, Chadha has framed this tension within the Ayodhyā conflict and the ASI's central role in conducting excavations and producing documentation in support of the Hindu litigants of the controversy. The judiciary instructed the ASI to investigate and produce a report within a relatively short period of time (Chadha 2016). This drew criticism from scholars and academics, both because the process was rushed, and for the uncorroborated findings.

INTACH publications

INTACH also produces an extensive repository of in-house publications covering case studies of both tangible and intangible heritage, including overviews of the built heritage of towns and cities such as Calcutta (Lal 2006), Kapurthala (INTACH, n.d.), and Lucknow (INTACH, n.d.). Site-specific studies include the likes of Mehrauli Archaeological Park[3] (Nalini Thakur), a 100-acre reserved forest land containing the World Heritage site of Qutub Minar; Leh Urban Conservation Project (Anuradha Chaturvedi), focusing on the ecology and culture of the town of Leh in the remote region of Ladakh; and the *ghats* in Mathura and Vrindavan, focusing on their restoration and continuing use (Ravindran 1990). INTACH has also published a compendium of good practices for urban heritage in India featuring topics such as heritage legislation, integration of conservation in urban renewal, revitalization of urban heritage through urban renewal, community participation, and heritage awareness (INTACH 2015). More recently, INTACH has started creating conservation briefs similar to the US's National Park Service publications as technical manuals for practitioners. This set includes a range of cultural resource types such as historic gardens (Priyaleen Singh), Mughal waterways (James Wescoat), historic materials including timber (Benny Kuriakose), and lime (Sangeeta Bais).

Other technical manuals published by INTACH include the *Guidelines for the Preparation of a Heritage Management Plan*, which details the development of such plans from visioning through implementation stages. Finally, the Intangible Cultural Heritage Division of INTACH has worked on documenting (including mapping) intangible heritage in various settings including Pondicherry, Varanasi, and Jandiala Guru in Punjab featured in this volume. Such efforts by INTACH are geared toward streamlining practice by providing tools to practitioners and sharing best practices.

One of the most significant contributions of INTACH has been their efforts at dramatically expanding the listing and documentation of built, natural, cultural, and intangible heritage relative to that of the ASI. This has, in turn, helped expand the mandate of heritage conservation, and create awareness about local heritage and facilitate action for protection. In the process, the organization has also developed criteria and methodology for listing. For example, its Delhi chapter has recorded 1,200 buildings of archaeological, historical and architectural importance, compared to 174 sites under the ASI's jurisdiction (Gupta, Jain, and INTACH 1999). The extensive listing program by INTACH Delhi has had the value of making municipal corporations more aware of their historic resources, a first step towards greater protection. However, in most cases, they have not yet taken significant additional steps to protect the listed heritage.

Academic publications

Outside of these ASI- and INTACH-focused (and generated) literature, there is a growing independent literature on heritage conservation in India. Some of this is a part of larger comparative volumes on heritage conservation within global, Asian, or South Asian contexts. These works have all brought attention to heritage conservation topics in India, but by their comparative nature cover a more limited scope of material within the Indian contexts. The recently published *Routledge Companion to Global Heritage Conservation* (Bharne and Sandmeier 2019) takes a global view on heritage conservation, particularly the challenges confronting practice, and features two chapters on India. One on Varanasi (also known as Banaras in the chapter) makes an argument for the culturally and historically important city to be a World Heritage Site, and for the integration of conservation and planning in the city (Singh and Rana 2019); another, on the preservation of the recent past in India, features Delhi's Pragati Maidan and Chandigarh (Chalana 2019). Additionally, scholarship on themes of heritage conservation in India can sometimes appear in published conference proceedings, like Mehrotra's work on Bombay's historic markets (Mehrotra 1998).

Works placing Indian conservation within a specifically Asian comparative context have covered a wider range of topics. Here scholars have explored relationships between modernity and the vernacular environment (Lim 2003; Lim and Tan 1998); urban heritage value and typology (Logan 2002), forms of modernity (King 2004), dynamism of the kinetic city (Mehrotra, 2007), and patterns of informality (Chalana and Hou 2016) – all having implications for broadening heritage conservation theory and practice. Logan's work in particular seeks to understand heritage policies, practices, attitudes, and global economic restructuring that threatens the

"Asian" character of the 14 featured cities, primarily in Southeast Asia, except one from India: Calcutta, known today as Kolkata (Ghosh 2002).

Heritage conservation topics in India can also appear in volumes on broader topics of urbanism and urban development in Asia; among the numerous examples are *The Emerging Asian City: Concomitant Urbanities & Urbanisms*, featuring the Viceroy's House in Delhi (Inam 2013), Chandigarh's Capitol (Bharne 2013), the Taj Mahal (Bharne 2013), and Mumbai (Chalana 2013). A recent publication has focused more directly on heritage conservation in the Asian context (Silva and Chapagain 2013), and features topics of the Hindu philosophy of conservation (Tom 2013); cultural heritage and sacred landscapes in Braj (Sinha 2013), and community-based approaches to heritage management in Leh (T. Sharma 2013). A recent volume by Stubbs and Thompson provides a comprehensive overview of preservation practice (organized by country) throughout Asia, with a section on India covering wide-ranging topics such as the religious underpinnings of restoration and conservation practice; colonial-era conservation principles and practice; conservation in the modern era; and challenges to cultural landscape conservation (Stubbs and Thomson 2016). This work intends to share local experiences and insights on best practices around cultural heritage management.

Another recent volume explores notions of authenticity in heritage conservation through European, South Asian, and East Asian lenses (Weiler and Gutschow 2017). The first part of the book focuses wholly on South Asia, featuring four chapters that raise issues of authenticity in the Indian context. The first essay explores the role of place in restoration and replication efforts in India (Menon 2017). The second essay traces the conservation journey of the iconic Humayun's Tomb in New Delhi (Nanda 2017), while the third chapter describes the community of Sompura from Western India, comprising generations of temple designers, builders, and master craftsmen whose trade has become part of the international discourse on authenticity (R. J. Vasavada 2017). The fourth chapter in this volume discusses the work of stonemasons and master builders in the context of the Nara Document to argue for the genuineness of Indian craftsmanship (Weiler 2017).

Heritage topics that place India in a specifically South Asian (as opposed to pan-Asian) context are rare, with one notable exception: Silva and Sinha's work on cultural landscapes, which covers a rich diversity of sites and associated traditional practices (Silva and Sinha 2016). A fifth of the book's contents focuses on India directly, including topics on community participation for regenerating historic neighbourhoods in Ahmedabad (Nayak 2016); application of cultural landscape methodology for managing linear cultural resources like the Grand Trunk Road (Chalana 2016); and management of the public realm as central to urban heritage conservation (Sinha 2016).

The final form of independent historic preservation scholarship (that is, works not produced by or about the ASI or INTACH) is that which is focused exclusively on India, without using an explicitly comparative lens. This list is expanding and covers a range of topics and geographies. An earlier anthology (Nilsson 1995a) explores different aspects of heritage conservation across cities in India: Ahmedabad (K. Jain 1995a), Lucknow (Llewellyn-Jones 1995), Jaisalmer (K. Jain 1995b), Jodhpur (K. Jain 1995b), Jaipur (Nilsson 1995b), Hyderabad (Shorey 1995), Junagadh (R. Vasavada 1995), Tranquebar (Hansen 1995), Pune (Nissen 1995), and Delhi (Gupta 1995). The chapters in this volume are well illustrated and offer a rich tapestry and historical and contemporary dimensions of heritage conservation. Since then, entire volumes focusing on heritage conservation in India are limited with the exception of Kulbhushan Jain's *Conserving Architecture*, which is richly illustrated and features different technical approaches to restoration projects incorporating personal experiences of professional conservationists (K. Jain 2017).

Several scholars and practitioners have published works on a variety of conservation topics across different Indian geographies. Chainani's work takes a broader view on heritage legislation and policies in the country (Chainani 2007a, 2007b, 1982), while Desai's work focuses on intersectionality of ethics and heritage in the context of a living historic city: Ahmedabad (Desai 2019). Many have worked on a range of history and heritage conservation topics in cities: Mehrotra, Dwivedi, and Lambah among others have published on Mumbai, focusing on naval heritage (Dwivedi et al. 2005); site and setting of a sacred tank in Malabar Hill (Mehrotra 2006); fort precedents exploration through walks (Dwivedi 1999), heritage streetscapes and citizen participation (Lambah 2012), and several others. Other towns and cities to have received attention on conservation topics include Lucknow (Nagpal and Sinha 2009; Sinha and Kant 2015; Sinha 2010; Krishna, n.d., 2014, 2012, 2017, 2016); Chandigarh (Chalana and Sprague 2013; Chalana 2016); Champaner-Pavagadh (Sinha 2004); Cochin (Fels 2011); Shekhawati (Lambah 2013); Mewar (S. Jain 2017); Banaras (Singh 2016; Singh and Rana 2017); Nabha, Punjab (Mubayi and Rai 2015); Shimla (S Sharma 2018); and small towns in the Gurgaon district (P. Munjal and Munjal 2017; P. G. Munjal 2017) among others. These engage a variety of heritage conservation topics and have expanded the mandate of the field in India along numerous dimensions. Some of these threads include patterns of indigenous modernity (Hosagrahar 2005), colonial and postcolonial geographies (Raju, Kumar, and Corbridge 2006), cultural landscapes (James Wescoat, Amita Sinha, and Priyaleen Singh), Indian cultural landscapes (Thakur 2012), the recent past (Chalana 2019; Chusid 2015), and vernacular heritage (Chalana 2009), among others. Several engage (to varying degrees) with the state of conservation practices in the context of regulations or lack thereof in India.

Our work builds on this existing literature in several ways. We have organized the volume into unique sections that address various facets of heritage conservation: understanding the policy and institutional frameworks applicable in India, analysing contemporary challenges, highlighting emerging trends, and promoting sustainable heritage conservation practices. The contributing authors to the volume are a mix of scholars and practitioners; we ensured to include both groups in sizable number to broaden the discourse. Finally, we have deliberately included topics related to education, art and artefacts, as well as intangible heritage, to engage the full range of heritage conservation.

Thematic outline of the book

This edited volume explores various themes in the broader context of the field of heritage conservation within India. The chapters have been organized into four thematic sections covering various facets of the profession in the postcolonial decades to demonstrate how the field has evolved and to highlight its continuing challenges.

PART I – *Developments in heritage conservation: institutions and programmes*

The first section, dealing with the developments in heritage conservation, comprises chapters that discuss how institutions and programs in India have evolved since independence. This is essential to understanding how heritage conservation was professionalized in India and highlights how practitioners (typically architects) address practical challenges in the field. This section builds on the introductory framework already presented that highlights the historical role of the ASI and INTACH in shaping the field. Sanyal's chapter (Chapter 1) shares insights gained from working with the ASI; it presents unique perspectives on the inner workings of the agency to clarify that ASI has the capacity to successfully carry out conservation projects without having to rely on outside consultants. He uses the example of Konarak and Sivasagar to discuss the empirical ways in which experts from the ASI conceptualise and carry out the conservation of sites under their management. Gupta's chapter (Chapter 2) traces the critical role played by INTACH in shaping heritage conservation policy and practice in India since the 1980s. He discusses INTACH's role in advocacy, outreach, education, and policy, and the significant and positive impact it has had on heritage conservation profession in India through the restoration and management of historic sites, and identification and listing of at-risk sites, among others.

Shikha Jain's chapter (Chapter 3) discusses how conservation planning of historic cities is critical for a more effective approach to urban development. She uses the cases of Jaipur and Ajmer to highlight city-level heritage management initiatives that simultaneously influenced government policies, as well as impacted phased implementation of large-scale urban conservation projects. The chapter illustrates how the sustainability of urban conservation projects is dependent on anchoring them to a city vision plan and an inclusive participatory approach. Krishna's chapter (Chapter 4) complements its predecessor and highlights how cities that do not undertake comprehensive heritage and urban planning, and are without local heritage policies, are able to still achieve successful conservation outcomes through citizen advocacy and activism. Using the example of Lucknow, she explores the work of a heritage advocate and activist in the city, who used civic engagement and public interest litigation (PIL) as tools to successfully compel the relevant agencies to carry out conservation. Tomlan's chapter (Chapter 5) addresses another important development in the field: programmes that promote heritage conservation among children through primary education and awareness in schools. He discusses the origins of early heritage education programs in the country and highlights contemporary programmes that are contributing to educating the youth. He focuses on a particular school program in Karnataka, near the World Heritage Site of Hampi, to argue for the benefits of heritage education in both public and private school systems for long-lasting positive impacts towards heritage conservation.

PART II – Critical challenges in heritage conservation

This section highlights various practice-oriented challenges faced by the field today. Rai and Manh (Chapter 6) follow the history of the summer palace of Maharaja Ranjit Singh in Amritsar: Rambagh Gardens and its gate. The authors examine the decade-long struggle to conserve the site's layered history and address contemporary challenges. Kothari investigates an urgent conservation issue (Chapter 7): the illicit trade of Indian art and artefacts, and places it within the current antiquities policy framework. She argues for better oversight, enforcement, and policy to prevent loss of movable heritage. Priya Jain focuses on another neglected aspect of heritage conservation in India: modern architecture. In her chapter (Chapter 8) she makes a case for the conservation of modern architecture, contrasting the challenges faced by Modernist sites designed by foreign architects with those created by Indian modernists. Jain argues that Indian modern heritage needs attention by both policymakers and citizens to ensure their protection. The chapter by Chalana and Vishwakarma (Chapter 9) looks at a small town in Himachal Pradesh, Chamba, to highlight how its cultural resources including vernacular architecture and cultural landscapes are left neglected, and threatened by unplanned development and natural hazards. They argue for the integration of heritage conservation and hazard mitigation to inform conservation approaches for Chamba and other such historic towns that remain hazard-prone.

PART III – Emerging trends in heritage conservation

Section three of the volume focuses on the emerging trends in the field of heritage conservation. Tipnis showcases her work from practice that focuses on at-risk urban and semi-urban heritage of the former colonial ports of Chinsurah, and Chandernagore, and the *havelis* of Old Delhi (Chapter 10). In each case, she draws from her professional work in devising sustainable solutions integrating digital technologies with heritage conservation. In her chapter, Bose details issues in conserving the threatened urban heritage of Kolkata's Chinatown (Chapter 11). This grassroots effort generated a cultural survey and architectural documentation as critical tools to safeguarding the city's unique minority neighbourhood: Chinatown.

Fels discusses vernacular mosques in Kerala within the context of their neighbourhoods in Cochin and Kozhikode (Chapter 12). She discusses the challenges faced by the current conservation practice in Kerala in maintaining the vernacular built environments and argues that conservation practice should protect the traditional mosques and homes for the residents. Mubayi's work (Chapter 13) focuses on the Thatheras of Jandiala Guru in Punjab, whose metalsmithing craft was inscribed on UNESCO's Intangible Cultural Heritage list in 2014. The chapter highlights the uniqueness of the manufacturing process, drawing from a complex range of traditional knowledge systems, and explores the process of recognition of the craft within the transforming context, and its continuing relevance.

PART IV – Sustainable approaches to heritage conservation

This section takes heritage conservation forward through discussions of sustainable approaches to heritage conservation projects. Sinha's chapter on Varanasi examines the complex landscape of *ghats* through an architectural, spatial, and cultural lens (Chapter 14). The *ghats* are continually under threat from development, flooding, tourism, and impacts of rituals and practices. She discusses sustainable approaches to resolving contemporary issues, arguing for using renewable energy sources to create a flexible and portable infrastructure for pilgrims and visitors. Wescoat discusses historical approaches to water-based Islamicate designs in northern and western India (Chapter 15). He explores qualitative approaches to conservation of water features in the Indian landscape and argues that the water features can assist in combating environmental issues, particularly related to drought and water runoff.

Singh explores the relationship between humans and nature through the exploration of historic gardens (Chapter 16). Her chapter illustrates the many ways that historic gardens can once again be brought to life and their meanings and messages recovered in order to connect with the communities that experience them. In the last chapter, Bhattacharya focuses on sustainable approaches to the conservation of intangible heritage in their original context (Chapter 17). She presents a community-led effort at promoting sustainable tourism integrated with cultural heritage in several rural communities in West Bengal through the Art for Life programme that empowers several struggling communities to revive traditional arts and crafts.

Finally, Krishna and Chalana conclude with a discussion on some of the continuing challenges the field of heritage conservation in India has to grapple with despite recent expansion in the multiple directions discussed in the chapters. One of the most pressing continuing challenges facing the field deals with the ASI's monument-centric approach, which has seen little evolution since the inception of the agency. Another major challenge seen most acutely in urban India is resolving the question of how heritage conservation can be integrated into planning and development initiatives and policy. Vernacular, rural, and intangible heritage continue to remain underappreciated, placing a large fraction of heritage at risk. Climate change, natural hazards, communal tensions, and global terrorism have also, in recent decades, impacted historic sites. Also, equity and social justice lenses are not fully engaged to make the field more inclusive. In the long term, conservation education would also need to evolve and expand to address these challenges and ensure that students and professionals in heritage conservation are trained in the myriad facets of the evolving field.

Notes

1 Full text available at http://asi.nic.in/wp-content/uploads/2018/06/new_6.pdf
2 For more on this see https://whc.unesco.org/archive/2015/whc15-20ga-9-en.pdf
3 The "Archaeological Park" methodology for the protection and management of heritage was detailed and contextualised for India by Nalini Thakur in the mid-1990s for the Mehrauli Archaeological Park in Delhi. The model has since been extended to other sites in the country.

References

Archaeological Survey of India. 1872. *Indian Antiquary.* https://catalog.hathitrust.org/Record/000501150

Bandyopadhyay, Bimal. 2004. *Buddhist Centres of Orissa: Lalitagiri, Ratnagiri, and Udayagiri.* New Delhi: Sundeep Prakashan.

Bandyopadhyay, Bimal. 2007. *Excavations at Udayagiri-2, 1997–2000. Memoirs of the Archaeological Survey of India; No. 100.* New Delhi: Director General, Archaeological Survey of India.

Bharne, Vinayak. 2011. "Le Corbusier's Ruin: The Changing Face of Chandigarh's Capitol." *Journal of Architectural Education* 64 (2). Wiley Online Library: 99–112. https://onlinelibrary.wiley.com/doi/full/10.1111/j.1531-314X.2010.01134.x

Bharne, Vinayak. 2013. "The Paradise between Two Worlds: Rereading Taj Mahal and Its Environs." In *The Emerging Asian City: Concomitant Urbanities and Urbanisms,* edited by Vinayak Bharne, 36–45. Abingdon: Routledge.

Bharne, Vinayak and Trudi Sandmeier, eds. 2019. *Routledge Companion to Global Heritage Conservation. Routledge Companion to Global Heritage Conservation.* Abingdon: Routledge. doi:10.4324/9781315659060.

Chadha, Ashish. 2007. *Performing Science, Producing Nation: Archaeology and the State in Post-Colonial India*. Palo Alto: Stanford University. http://search.proque st.com/docview/304807625?accountid=14169

Chadha, Ashish. 2016. "Commentary: Archaeological Survey of India and the Science of Postcolonial Archaeology." In *Handbook of Postcolonial Archaeology*, edited by Jane Lydon and Uzma Z. Rizvi, 227–234. Abingdon: Routledge. https ://www.routledgehandbooks.com/doi/10.4324/9781315427690

Chainani, Shyam. 1982. "The Role of Environmental Groups: A Bombay Experience." *India International Centre Quarterly* 9 (3): 268–281.

Chainani, Shyam. 2007a. *Heritage & Environment: An Indian Diary*. Mumbai: Urban Design Research Institute.

Chainani, Shyam. 2007b. *Heritage Conservation, Legislative and Organisational Policies for India*. New Delhi: INTACH.

Chalana, Manish. 2009. "Of Mills and Malls: The Future of Urban Industrial Heritage in Neoliberal Mumbai." *Future Anterior* 9 (1). University of Minnesota Press: a-15. https://muse.jhu.edu/journals/future_anterior/v009/9.1. chalana.html

Chalana, Manish. 2013. "Making Way for a Global Metropolis: Mumbai's Rapidly Transforming Informal Sector." In *The Emerging Asian City: Concomitant Urbanities and Urbanisms*, edited by Vinayak Bharne, 193–202. Abingdon: Routledge.

Chalana, Manish. 2016. "'All the World Going and Coming': The Past and Future of the Grand Trunk Road in Pubjab, India." In *Cultural Landscapes of South Asia: Studies in Heritage Conservation and Management*, edited by Kapila D. Silva and Amita Sinha. Abingdon: Routledge, 92–110.

Chalana, Manish. 2019. "The Future of the Recent Past: Challenges Facing Modern Heritage from the Postcolonial Decades in India." In *Routledge Companion to Global Heritage Conservation*, edited by Vinayak Bharne and Trudi Sandmeier, 477–486. Abingdon: Routledge.

Chalana, Manish and Jeffrey Hou, eds. 2016. *Messy Urbanism Understanding the 'Other' Cities of Asia*. Hong Kong: Hong Kong University Press. http://www. hkupress.org.

Chalana, Manish and Tyler S. Sprague. 2013. "Beyond Le Corbusier and the Modernist City: Reframing Chandigarh's 'World Heritage' Legacy." *Planning Perspectives* 28 (2): 199–222. doi:10.1080/02665433.2013.737709.

Chusid, Jeffrey Mark. 2015. "The India International Centre of Joseph Allen Stein: A Story of Cold War Politics and the Preservation of a Modern Monument." *Journal of Architectural Conservation* 21 (2): 71–84. doi:10.1080/13556207.2015.1083292.

Desai, Jigna. 2019. *Equity in Heritage Conservation. Equity in Heritage Conservation*. Routledge. doi:10.4324/9780429468735.

Dwivedi, Sharada. 1999. *Fort Walks: Around Bombay's Fort Area*. Bombay: Eminence Designs.

Dwivedi, Sharada, Rahul Mehrotra, Abha Narain Lambah, Jehangir Sorabjee and Chirodeep Chaudhuri. 2005. *A City's Legacy: The Indian Navy's Heritage in Mumbai*. Mumbai: Eminence Designs.

Fels, Patricia Tusa. 2011. *Mosques of Cochin*. Ahmedabad: Mapin Publishing.

Ghosh, Santosh. 2002. "Calcutta, India: Heritage and Government Priorities." In *The Disappearing Asian Cities*, edited by William Stewart Logan, 105–123. Hong Kong: Oxford University Press.

Gupta, Narayani. 1995. "The Cities of Delhi." In *Aspects of Conservation in Urban India*, edited by Sten Nilsson, 215–239. Lund: Lund University Press.

Gupta, Narayani, O.P. Jain and INTACH. 1999. *Built Heritage: A Listing*. New Delhi: INTACH Press.

Hansen, Hans Munk. 1995. "Tranquebar." In *Aspects of Conservation in Urban India*, edited by Sten Nilsson, 155–180. Lund: Lund University Press.

Hosagrahar, Jyoti. 2005. *Indigenous Modernities : Negotiating Architecture and Urbanism*. London: Routledge.

Inam, Aseem. 2013. "Tensions Manifested: Reading the Viceroy's House." In *The Emerging Asian City: Concomitant Urbanities and Urbanisms*, edited by Vinayak Bharne, 99–109. Abingdon: Routledge.

INTACH. n.d. *Kapurthala, a Study of Genteel Elegance*. New Delhi: INTACH Press.

INTACH. n.d. *Lucknow, a Treasure*. New Delhi: INTACH Press.

INTACH. 2015. *Compendium of Good Practices: Urban Heritage in Indian Cities*. New Delhi: INTACH Press.

Jain, A.K. 2010. "Interface between Traditional Urbanism and the Legislative Framework." In *New Architecture and Urbanism: Development of Indian Traditions*, edited by Deependra Prashad, 298–303. Newcastle upon Tyne: Cambridge Scholars Publishing.

Jain, Kulbhushan. 1995a. "Ahmedabad--a City and Its Conservation." In *Aspects of Conservation in Urban India*, edited by Sten Nilsson, 25–54. Lund: Lund University Press.

Jain, Kulbhushan. 1995b. "Jaisalmer and the Effects of Tourism." In *Aspects of Conservation in Urban India*, edited by Sten Nilsson, 79–98. Lund: Lund University Press.

Jain, Kulbhushan, ed. 2017. *Conserving Architecture*. Ahmedabad: AADI Centre.

Jain, Shikha. 2017. *Living Heritage of Mewar: Architecture of the City Palace, Udaipur*. Ahmedabad: Mapin Publishing.

Johnson-Roehr, Susan. 2008. "The Archaeological Survey of India and Communal Violence in Post-Independence India." *International Journal of Heritage Studies* 14 (6): 506–523. doi:10.1080/13527250802503266.

Keay, John. 2011. *To Cherish and Conserve: The Early Years of the Archaeological Survey of India*. New Delhi: National Culture Fund, Archaeological Survey of India.

King, Anthony. 2004. *Spaces of Global Cultures: Architecture, Urbanism, Identity*. London: Routledge.

Krishna, Ashima. n.d. "Politics, the Public, and Urban Conservation in Lucknow, India." In *Untitled*, forthcoming.

Krishna, Ashima. 2012. "The Business of Heritage." *Context: Built, Living and Natural* 9 (2): 13–20. http://search.proquest.com/docview/1372088072?pq-o rigsite=summon&accountid=10226

Krishna, Ashima. 2014. *The Urban Heritage Management Paradigm: Challenges from Lucknow, An Emerging Indian City*. Cornell University. http://ecommons .library.cornell.edu/handle/1813/36044

Krishna, Ashima. 2016. "The Catalysts for Urban Conservation in Indian Cities: Economics, Politics, and Public Advocacy in Lucknow." *Journal of the American Planning Association* 4363 (January): 1–4. doi:10.1080/01944363.2015.1132390.

Krishna, Ashima. 2017. "From Alliance to Dissonance: Two Centuries of Local Archaeology and Conservation in Indian Cities. The Case of Lucknow, India."

In *Urban Archaeology, Municipal Government and Local Planning*, edited by Sherene Baugher, Douglas Appler and William Ross. Springer International Publishing AG. https://link.springer.com/chapter/10.1007/978-3-319-55490-7_9

Lal, Nilina Deb. 2006. *Calcutta: Built Heritage Today: An INTACH Guide*. Calcutta: INTACH Calcutta Regional Chapter.

Lambah, Abha Narain. 2012. "Mumbai: Historic Preservation by Citizens." In *Cultures and Globalization: Cities, Cultural Policy and Governance*, edited by Helmut K Anheier and Yudhishthir Raj Isar. London: SAGE Publications. 251–256.

Lambah, Abha Narain. 2013. *Shekhawati, Havelis of the Merchant Princes*. v. 65, No. 1. Mumbai: Marg Publications.

Lim, William Siew Wai. 2003. *Alternative (Post) Modernity: An Asian Perspective*. Singapore: Select Books.

Lim, William Siew Wai and Hock Beng Tan. 1998. *Contemporary Vernacular: Evoking Traditions in Asian Architecture*. Singapore: Select Books. https://ezp.lib.unimelb.edu.au/login?url=https://search.http://ebscohost.com/login.aspx?direct=true&db=edshlc&AN=edshlc.007454873-5&scope=site

Llewellyn-Jones, Rosie. 1995. "Conservation in Lucknow." In *Aspects of Conservation in Urban India*, edited by Sten Nilsson, 55–70. Lund: Lund University Press.

Logan, William Stewart. 2002. *The Disappearing "Asian" City: Protecting Asia's Urban Heritage in a Globalizing World*. Oxford University Press. https://books.google.com/books/about/The_Disappearing_Asian_City.html?id=BUl5QgACAAJ&pgis=1

Mani, B.R. 2012. *Catalogue of Terracotta Figurines from Rajghat Excavations (1940). Memoirs of the Archaeological Survey of India; No. 108*. New Delhi: Director General, Archaeological Survey of India.

Marshall, John Hubert. 1923. *Conservation Manual. A Handbook for the Use of Archaeological Officers and Others Entrusted with the Care of Ancient Monuments*. Calcutta: Superintendent Government Printing India.

Mehrotra, Rahul. 1998. "Bazaars in Victorian Arcades: Conserving Bombay's Historic Core." In *City Space + Globalization: An International Perspective*, edited by Hemalata Dandekar. Ann Arbor: University of Michigan. 46–53.

Mehrotra, Rahul. 2006. *Banganga: Sacred Tank on Malabar Hill*. Rev. ed. Girgaum: Eminence Designs.

Mehrotra, Rahul. 2007. "Conservation and Change: Questions for Conservation Education in Urban India." *Built Environment* 33 (3): 342–356. https://doi.org/10.2148/benv.33.3.342

Menon, A.G. Krishna. 2017. "The Idea of Place in the Practice of Restoration and Replication in India." In *Authenticity in Architectural Heritage Conservation: Discourses, Opinions, Experiences in Europe, South and East Asia*, edited by Katharina Weiler and Niels Gutschow, 1st ed. Switzerland: Springer International Publishing. 87–92.

Mubayi, Yaaminey and Gurmeet S. Rai. 2015. "Cultural Heritage as a Driver for Integrated Development in Punjab: The Case of Nabha." *New Architecture and Urbanism: Development of Indian Traditions* 1: 47–53. doi:10.5848/csp.1892.00006.

Munjal, Parul G. 2017. "Construction of Heritage: Small and Medium Towns of Gurgaon District." *Journal of Heritage Management* 1 (2): 98–125. doi:10.1177/2455929616682079.

Munjal, Parul G. and Sandeep Munjal. 2017. "Built Heritage in Small Towns a Unique Tourism Opportunity: Case of Shiv Kund, Sohna." *Journal of Services Research* 17 (2): 17–40.

Nagpal, Swati and Amita Sinha. 2009. "The Gomti Riverfront in Lucknow, India: Revitalization of a Cultural Heritage Landscape." *Journal of Urban Design* 14 (4): 489–506. doi:10.1080/13574800903264838.

Nanda, Ratish. 2017. "Humayun's Tomb: Conservation and Restoration." In *Authenticity in Architectural Heritage Conservation: Discourses, Opinions, Experiences in Europe, South and East Asia*, edited by Katharina Weiler and Niels Gutschow, 93–114. Switzerland: Springer International Publishing.

Nayak, Debashish. 2016. "Getting the City Back to Its People: Conservation and Management of Historic Ahmedabad, India." In *Cultural Landscapes of South Asia: Studies in Heritage Conservation and Management*, edited by Kapila D. Silva and Amita Sinha. Abingdon: Routledge, 211–226.

Nilsson, Sten. 1995a. *Aspects of Conservation in Urban India (Studies in Art History)*. Lund: Lund University Press.

Nilsson, Sten. 1995b. "Jaipur--Reflections of a Celestial Order." In *Aspects of Conservation in Urban India*, edited by Sten Nilsson, 107–128. Lund: Lund University Press.

Nissen, Foy. 1995. "Pune Cantonment." In *Aspects of Conservation in Urban India*, edited by Sten Nilsson, 181–214. Lund: Lund University Press.

Pant, Dhirendra. 2012. *Care and Administration of Heritage Monuments in India, 1784–1904*. New Delhi: Aryan Books International.

Patnaik, Jeeban Kumar. 2016. *Excavations at Lalitagiri (1985–1991), 1st ed. Memoirs of the Archaeological Survey of India; No. 112*. New Delhi: Director General, Archaeological Survey of India.

Raju, Saraswati, M. Satish Kumar and Stuart Corbridge. 2006. *Colonial and Post-Colonial Geographies of India*. Thousand Oaks: SAGE.

Ravindran, K.T. 1990. *The Ghats of Mathura and Vrindavan, Proposals for Restoration*. INTACH Cultural Heritage Case Studies. New Delhi: Indian National Trust for Art & Cultural Heritage.

Sengupta, Gautam, Abha Narain Lambah and Archæological Survey of India. 2012. *Custodians of the Past: 150 Years of the Archaeological Survey of India*. New Delhi: Archaeological Survey of India, Ministry of Culture, Government of India.

Sengupta, Indra. 2013. "A Conservation Code for the Colony: John Marshall's Conservation Manual and Monument Preservation Between India and Europe." In *'Archaeologizing' Heritage*, edited by Michael Falser and Monica Juneja, 1st ed., 21–38. Heidelberg: Springer. doi:10.1007/978-3-642-35870-8.

Sharma, Jahnwij. 2012. "Conservation of Monuments in India: A 150-Year Perspective." In *Custodians of the Past: 150 Years of the Archaeological Survey of India*, edited by Gautam Sengupta and Abha Narain Lambah, 120–143. New Delhi: Archaeological Survey of India, Ministry of Culture, Government of India.

Sharma, Tara. 2013. "A Community-Based Approach to Heritage Management from Ladakh, India." In *Asian Heritage Management: Contexts, Concerns, and Prospects*, edited by Kapila D. Silva and Neel Kamal Chapagain, 271–284. Abingdon: Routledge.

Shorey, S.P. 1995. "Hyderabad and the Future of Lad Bazaar." In *Aspects of Conservation in Urban India*, edited by Sten Nilsson, 129–140. Lund: Lund University Press.

Silva, Kapila D. and Neel Kamal Chapagain, eds. 2013. *Asian Heritage Management*. Asian Heritage Management. Abingdon: Routledge. doi:10.4324/9780203066591.

Silva, Kapila D. and Amita Sinha, eds. 2016. *Cultural Landscapes of South Asia: Studies in Heritage Conservation and Management*. Cultural Landscapes of South Asia: Studies in Heritage Conservation and Management. Routledge. doi:10.4324/9781315670041.

Singh, Rana P.B. 2016. "Urban Heritage and Planning in India: A Study of Banaras." In *Spatial Diversity and Dynamics in Resources and Urban Development: Volume II: Urban Development*, edited by Ashok K. Dutt, Allen G. Noble, Frank G. Costa, Rajiv R. Thakur and S.K. Thakur, 423–449. Dordrecht: Springer Science + Business. doi:10.1007/978-94-017-9786-3.

Singh, Rana P.B. and Pravin S. Rana. 2017. "Varanasi : Heritage Zones and Its Designation in UNESCO ' s World Heritage Properties." *Kashi Journal of Social Sciences* 7 (December): 201–218.

Singh, Rana P.B. and Pravin S. Rana. 2019. "Visioning Cultural Heritage and Planning: Banaras, the Cultural Capital of India." In *Routledge Companion to Global Heritage Conservation*, edited by Vinayak Bharne and Trudi Sandmeier, 152–175. Abingdon: Routledge.

Sinha, Amita. 2004. "Champaner-Pavagadh Archaeological Park: A Design Approach." *International Journal of Heritage Studies* 10 (2): 117–128. doi:10.1080/13527250410001692859.

Sinha, Amita. 2010. "Colonial and Post-Colonial Memorial Parks in Lucknow, India: Shifting Ideologies and Changing Aesthetics." *Journal of Landscape Architecture* 2010 (10): 60–71. http://www.tandfonline.com/doi/abs/10.1080/18626033.2010.9723439

Sinha, Amita. 2013. "Cultural Heritage and Sacred Landscapes of South Asia: Reclamation of Govardhan in Braj, India." In *Asian Heritage Management: Contexts, Concerns, and Prospects*, edited by Kapila D. Silva and Neel Kamal Chapagain, 176–188. Abingdon: Routledge.

Sinha, Amita. 2016. "The Public Realm of Heritage Sites in India: Sustainable Approaches towards Planning and Management." In *Cultural Landscapes of South Asia: Studies in Heritage Conservation and Management*, edited by Kapila D. Silva and Amita Sinha. Abingdon: Routledge, 227–249.

Sinha, Amita and Rajat Kant. 2015. "Mayawati and Memorial Parks in Lucknow, India: Landscapes of Empowerment." *Studies in the History of Gardens & Designed Landscapes* 35 (1): 43–58. doi:10.1080/14601176.2014.928490.

Stubbs, John H. and Robert G. Thomson. 2016. *Architectural Conservation in Asia: National Experiences and Practice*. Architectural Conservation in Asia: National Experiences and Practice. Routledge. https://doi.org/10.4324/9781315683447

Thakur, Nalini. 2002. "Champaner Pavagadh Archaeological Park—An Integrated Approach with Comprehensive Protection and Management." Unpublished work.

Thakur, Nalini. 2012. "The Indian Cultural Landscape: Protecting and Managing the Physical to the Metaphysical Values." In *Managing Cultural Landscapes*, edited by Ken Taylor and Jane L. Lennon, 154–172. Abingdon: Routledge.

Tom, Binumol. 2013. "Jiirnnoddhharana: The Hindu Philosophy of Conservation." In *Asian Heritage Management: Contexts, Concerns, and Prospects*, edited by Kapila D Silva and Neel Kamal Chapagain, 35–48. Abingdon: Routledge.

Vasavada, Rabindra J. 2017. "Sompura: Traditional Master Builders of Western India." In *Authenticity in Architectural Heritage Conservation: Discourses, Opinions, Experiences in Europe, South and East Asia*, edited by Katharina Weiler and Niels Gutschow, 115–126. Switzerland: Springer International Publishing.

Vasavada, Ravindra. 1995. "Nawabi Architecture in Junagadh." In *Aspects of Conservation in Urban India*, edited by Sten Nilsson, 141–154. Lund: Lund University Press.

Weiler, Katharina. 2017. "Contested Evaluations: Authenticity and the 'Living Traditions' of Master Builders and Stonemasons in India." In *Authenticity in Architectural Heritage Conservation: Discourses, Opinions, Experiences in Europe, South and East Asia*, edited by Katharina Weiler and Niels Gutschow, 127–168. Switzerland: Springer International Publishing.

Weiler, Katharina and Niels Gutschow. 2017. "Authenticity in Architectural Heritage Conservation: Discourses, Opinions, Experiences in Europe, South and East Asia." *Transcultural Research – Heidelberg Studies on Asia and Europe in a Global Context*. Switzerland: Springer International Publishing.

Part I

Developments in heritage conservation: Institutions and programs

1 The evolving role of India's foremost heritage custodian

The Archaeological Survey of India

Saptarshi Sanyal

Introduction

Both students and independent practitioners of heritage conservation in India have, for long, viewed the policies and/or actions of the Archaeological Survey of India (ASI) rather cautiously. This stemmed from a general rhetoric in Indian professional circles about the agency being intellectually and administratively ill-equipped to carry out appropriate conservation practices. Public opinion about the organisation's high-handedness and resistance to change also prevailed. Coming from a background in conservation training, this outlook persisted in the initial phase of my tenure at the ASI as an in-house conservation architect from 2010 until 2013. In this chapter I draw on my experience of working within the ASI and how some of the negative perceptions about the agency were consequently challenged. Through the forthcoming discussions, I demonstrate how the ASI has the capacity to evolve its professional practices by recognising the value of a *protracted* engagement with archaeological sites and structures over time.

The ASI has had separate organisational wings for archaeology and conservation – a legacy of the British era where the latter field was dealt in a technocratic manner. Consequently, history and scholarship about material remains (archaeology) and their physical conservation were seen as unrelated endeavours. Even today, conservation work is entrusted to a separate section under 'engineering'. This organisational structure has a negative impact on conservation sites, as on-site investigations and technical intervention tend to be divided by rubrics of the humanities and applied sciences. In this work I demonstrate how integration of research and on-site interventions can transform practice and challenge certain stereotypes about the ASI. I also argue that the agency is capable of carrying out conservation of sites internally in a cost-effective manner without having to rely on private consultants.

Legacy and evolution of the ASI

Established in 1861 in British India, the ASI is one of the oldest technical arms of the present Government of India. It has a legacy of over a century and half of exploration, identification, and notification of heritage sites in the country. A substantial portion of India's material heritage has therefore historically been synonymous with the ASI. With the Constitution of India recognising built heritage as a significant public good (Ministry of Law and Justice 2015, 24; 325–328)[1], the ASI was considered instrumental in implementing this national responsibility. Consequently, in independent India, it was formally elevated to the status of a premier organisation under the Ministry of Culture. Protection, management, conservation, and maintenance of built heritage have been among the ASI's prime concerns, besides regulation of archaeological activities and antiquities. As of 2020, the organisation protects and maintains more than 3500 sites in the country (ASI 2011). It is also the nodal agency and custodian of 26 of India's 30 (cultural)[2] World Heritage Sites (WHC 2020). In addition, it carries out 'deposit works'[3] for heritage properties under other institutions, both within and outside the country (ASI 2006).

In the post-colonial context, the ASI has been extensively, and legitimately so, critiqued for perpetuating outdated 'colonial' paradigms in many areas. The criticisms pertain to ad hoc problem-solving and approaches to repair and maintenance, often uninformed by research and by previous interventions, which could lead to misguided efforts. The ASI's legal protection framework and approaches to conservation have also been discussed critically both nationally and internationally (Allchin 1978, 747–752). This is primarily with regard to the mode in which monuments or sites are maintained as 'artefacts', often disconnected from their living contexts while alienating local populations: a major social concern. The agency is also criticised in the public eye for outsourcing work to private consultants, which raises questions about long-term accountability, sustainability, and monitoring of actions on sites (Anand 2015; Casey 1995).

It is in this context that an approach of critical and continuous engagement with sites within the ASI constitutes an important methodological departure. This is presented in further detail in the sections that follow, through two distinct cases. It may be noted that such a shift is not unprecedented. The Integrated Management Plan for the Hampi World Heritage Site is a relevant recent example, which endeavoured to bring several public and private sector agencies together under a joint long-term programme for spatial heritage management within a legal framework (Thakur 2007, 32–36). The ASI was a major stakeholder in this effort, which highlighted a significant step towards an emerging conservation paradigm in post-colonial India. Such a paradigm, however, is not without challenges in the ASI's resource-starved bureaucratic context. The problems of not fully assessing a site's values, attributes and condition has undermined the potential of the ASI's work for decades. Many sites lack comprehensive documentation and current scholarly perspectives are not engaged. As a result, actions that are detrimental to a site's history and continuity inform interventions. Therefore, use of historical and contemporary research sources to inform appropriate on-site intervention strategies is critical. First, it adds to extant knowledge about the site. Second, it generates a continuous record of historic and recent interventions to guide future action, completing a 'feedback loop' on the performance of conservation action, as well as the future monitoring of the sites and structures.

The Sun Temple at Konarak

Located in close proximity to the shores of the Bay of Bengal, the Sun Temple falls within the historic Kalinga region, which now roughly constitutes the eastern state of Odisha (formerly Orissa) in India. The site represents a culmination of several centuries of temple building in the region (Donaldson 2005; Starza-Majeswski 1971, 135). It is one of the largest and most elaborate building projects among contemporary temples under the Ganga rulers demonstrating both continuity and innovation within canonical traditions (Behera 1996, 128). It is also one of the earliest World Heritage Sites to be inscribed from India, in 1984, based on three of the six cultural criteria of UNESCO. These include criterion I: representation of a unique artistic achievement; criterion III: outstanding testimony to the thirteenth-century kingdom of Odisha; criterion VI: link in the diffusion of the Tantric cult of *Surya* (Sun) worship (WHC 2018).

The site was in partial ruin even at the time of the UNESCO inscription. Its structures include an audience hall (or *jagamohana*), the base of the main sanctuary and the roofless dance hall (or *natmandir*) along with remains of two older temples within its walled premises. Along with its architectural form and scale, criterion I is also justified by all structures exhibiting some of the most outstanding examples of sculptural art in India (ICOMOS 1984) (Figure 1.1).

This broad knowledge on what survives at the site was complemented by the work of the ASI team from 2010–12. The team, including the author, comprising archaeologists, conservators, and engineers, focused their work on the structural and fabric assessment for potential conservation interventions in the audience hall. In this case, a detailed current examination of the structure was compared with relevant historical evidence. These records included a palm leaf manuscript, *Madala Panji*, and the book of accounts, *Byaya Chakada*, which described the materials, construction process, and the structure's physical state over time, respectively. (Local historians contributed in the process by examining the indigenous cultural records.) The team also critically looked into the authenticity of these records, which has been questioned by some historians (Behera 1996, 178–180).[4] This helped to isolate the credible sources for understanding the physical state of the structure over time. Important information about the site chronicled by Emperor Akbar's historian, Abu-al-Fazl, whose sixteenth-century account describes with dimensions, the whole temple, was especially relevant (Jarrett and Sarkar 1949, 140–141). Its various states of disrepair were described by

Figure 1.1 Remains of sanctuary and audience hall (*jagmohana*) of the Temple from the west, showing the form and scale of the structure (left); view of the Dance Hall from the east (right). Source: Author, 2010

the Marathas' Baba Brahmachari in 1759; British explorer Andrew Sterling in 1825; and engineer Bishan Swarup in 1910 (Sharma and Sanyal 2010, 15–18). Swarup was incidentally involved in the largest and most lasting intervention in the audience hall, the filling of its interior; his drawings of civil works greatly informed specific approaches within the project.

In addition to these texts, images of the site from the nineteenth century, held in the British Library's India Office Collection provided significant evidence on the past conditions of the site. Apart from the exterior views painted by the East India Company's artists in 1809 and 1820, and James Fergusson's detailed lithograph of 1837, William George Stephen's 1812 painting comprised key sources for the studies (Bose 1926, 150–152). Stephen's painting is, until today, the only available visual representation of the interior of the audience hall's structure.

These visuals were also corroborated by two important publications that provided detailed overviews of interventions carried out by both the British government's District Commissioner in Puri, and, subsequently, the ASI in independent India (Mitra 1968, 13–23; Chauley 1997). For a deeper appreciation of technical issues, various experts' recommendations over time were consulted. Some important ones included Sir Bernard Feilden's (1987) and Prof. Giorgio Croci's (1995) observations about the structural condition of the site. In addition, the ASI's Science Branch reports (2010) explained the structure and materials' physical and chemical properties. Apart from this scientific information, early twentieth-century images, from both the Asiatic Society and National Library Archives in Kolkata (Fergusson 1860; Mukherjee 1910), helped to corroborate the various states of conservation of the fabric and structure. All this information, collectively, supplied invaluable data for the project.

The project's approach combined several methods: physical examination, critical review of sources of documentation, and previous research and site inspections. The team took into account relevant evidence, instead of purely aiming for cosmetic repairs of the 'visible'. This process of research also developed a shared understanding of conservation priorities within the organisation. It brought about an informed approach that helped to delineate focussed goals for the site's conservation.

Ahom monuments at Sivasagar

Within the wide river valley landscape of Dikhow and its tributary stream, Namdang, Sivasagar's architectural and archaeological remains are a lasting legacy of its Ahom rulers. Located in the upper Assam region in northeast India, Sivasagar is now a district that borders the present state of Nagaland. Historians trace the origins of the Ahoms to travelling clans from geographical areas east of the Brahmaputra Valley, narrating their political control over much of the northeastern region of India for nearly six hundred years, beginning in 1228 CE (Gait 1906, 67–229). The region is significant for retaining a large variety of building types from different time periods, embodying a variety of building traditions in brick, stone, and lime (Phukan 1973, 131–133). Both continuities and mutations of these building traditions also comprise an important testimony to the various cultural exchanges that the Ahoms partook in since the sixteenth century with Brahmanical as well as Indo-Islamic ideologies (Gogoi 2011, 207).

Representing the ASI's Regional Directorate (East), and supported by its Guwahati Circle's conservation personnel, photographers, and archaeologists, the author carried out a literature survey and preliminary reconnaissance at Sivasagar. There was a large gap between existing cultural histories for Sivasagar and available site specific information. Gait's monograph refers to and describes the contents of *Buranjis*, which are indigenous historical chronicles containing meticulous records of different royal administrations in Sivasagar and their activities. Seventeen of these *Buranjis* in both the vernacular Ahom and Assamese languages survived at his time of writing in 1906. They have since been indispensable sources for authoritative scholarship by later local historians including H.K. Barpujari (1990).

The historical texts notwithstanding, unlike with the Sun Temple, no specific physical descriptions of the sites accompanied overall historical accounts on the Ahoms. Furthermore, except for a few rudimentary drawings with the Guwahati Circle of ASI, originally prepared in the early twentieth century[5] and corresponding images, there was no survey drawing available for the region of Sivasagar as a whole, as well as the outlying structures scattered in the Dikhow Valley region, despite their probable historical significance.

The interventions at the Ahom monuments in Sivasagar were originally, in 2010, conceived as a small beautification project funded by the Oil and Natural Gas Corporation (ONGC) for a 'facelift' of these areas to bring them into Assam's tourism network. Given a dearth of site-specific information, the ASI consequently negotiated the objectives with ONGC into developing a conservation management plan. The lack of information on the Ahom sites made it imperative to gather knowledge about the site through detailed architectural documentation, which became one of the priorities of the conservation approach. Consequently, a plan was formulated in 2010 and documentation was carried out from 2011–12 (ASI RD – ER 2011, 10–22). The headquarters of the ASI in New Delhi supported this course of action, resulting in a consultative process with various stakeholders, including ONGC, state and local governments, residents around the sites, and visitors.

The plan's first stage included an area mapping exercise and selection of relevant sites through historical research. Due to limited resources at the offices of the ASI in the northeast region of India, four pilot sites were selected to cover important representative periods of Ahom history. The oldest of these dates to the thirteenth century CE: *Maidams* – brick burial chambers embedded in earth berm structures that testify to the ancient ancestor worship traditions within Ahom culture. The second, a historic palace at Garghaon, originally erected in the sixteenth century, is testimony to layers of architectural alterations in the Ahom period until the eighteenth century and the adaptation of varied building techniques to Ahom royal architecture. Lastly, two eighteenth-century sites, Talatal Ghar and Rang Ghar, a palace and an entertainment pavilion respectively, were included as representation of the last capital of the Ahoms, Rangpur (Barpujari 1990) (Figure 1.2).

For Sivasagar, the ASI's original scope of work, as solicited by the ONGC, was to select sites for beautification and advise on selection of contractors for tourism development. In this case, the reconnaissance and initial research carried out by the ASI team revealed very different priorities. It was also deemed necessary that instead of outsourcing the work to private contractors, ASI would lead this exercise in-house, marking a significant shift from its conventional practice. With administrative support from the ASI Headquarters, the plan's development and deployment was undertaken in a decentralised way to ensure capacity building at ASI's local offices, quality management, as well as accountability.

Approaching the process of conservation

As both Konarak and Sivasagar demonstrate, data collection before any intervention helped to select the most relevant information and issues for conservation in an efficient manner and assisted in making precise decisions towards planning and action for conservation. The experience was heuristic because it involved focussed and exploratory engagement with the site over time,

Figure 1.2 The nature of structures selected for pilot documentation in Sivasagar – (clockwise) Partly excavated 'Maidam' at Charaideo showing brick chamber embedded within the earth berm structure; the pavilion Rang Ghar at the last capital of the Ahoms; Garghaon Palace. Source: Author, 2010

along with an organised approach to observation and learning (Bach 2002, 92–96). The intent to understand the structure, fabric, and historic landscape was more important than to carry out any preconceived action – at Konarak through cosmetic repairs that would conceal structural and fabric damage, and at Sivasagar through beautification of the sites for tourism. The rigour and discipline of immersive practice at the sites helped in improving the capacity of staff by deepening their sensibilities and sensitivity to the sites. The work, in this sense, helped an enhanced understanding of nuances of the local cultures and their connection to the sites, and the 'language' relevant for carrying out actions on-site, akin to developing a 'hermeneutic' sensibility (Malbon 1983, 221–222) related to the historic built environment. It interpreted and incorporated local and indigenous knowledge within the technical approach. Exposition by regional scholars as well as local representatives, who had a wealth of observations, was critical to guide site investigations in the conservation process.

Structure and fabric of the Sun Temple

In the course of the conservation process, two major issues related to the Sun Temple came to the fore. The first related to uncertainty about the surviving audience hall's structural safety.[6] In both its form and fabric, this structure dominates the ruins.[7] Weighing about 46,000 tonnes (Feilden 1987, 5–7), the walls are constructed of three-leaf dry masonry with laterite stone for core and khondalite sandstone for external and sculptural surfaces.[8] Anticipating structural duress due to falling stones recorded throughout the nineteenth century, the structure's interior was filled with sand and its openings sealed during the British administration from 1903–08. The supervising engineer, Bishan Swarup, noted in his records that sand was contained within a dry stone casing: about five metres in thickness and fifteen metres in height from the plinth. He indicated that the total volume of materials, which included both the stone casing and sand, weighed about two thousand tonnes (1980, 10–12).

Given this historical intervention, it was important to diagnose any symptoms of distress displayed by the structure. Due to lack of access to the interior, very little was known about the integrity of the structure for nearly a century. The ASI team relied on Croci's assumptions that the inner layers of corbelled masonry were not contributing to the overall stability of the structure (Croci 1987, 94). The team's inspection of vents (used originally to load sand) that were left open allowed for passage of air, which would only be possible if the space was hollow from the inside at or below the observation level. This led to the inference that the sand filling rested below the height of the vents. From available architectural documentation, Swarup's drawings, and mathematical assumption of the angle of repose of sand, it was assumed that the sand now rested in a pyramidal form lower than the soffit of the corbelled roof. Hence, it was not serving its intended purpose of preventing the loose masonry stones from falling into the interior as shown in Figure 1.3 (Sanyal and Sharma 2010, 45).

The next task involved determining the impact of the sinking sand on the structure. Vertical cracks at certain locations in the lower part of the structure indicated the stress of additional weight exerting load on the walls. Over several months, the breakage in tell-tale markings[9] affixed to monitor these cracks indicated their expansion over time, affirming this theory. Given the depressed condition of sand, it was understood that the thrust on the walls was lateral. The priority was therefore to facilitate access to the interior to study its condition, and the mechanical effect of the sand filling on the structure. It was proposed that a passage into the interior be created through the original opening to execute the filling. This would allow the team to analyse the interior situation, develop numerical models of structural stresses, and carry out requisite structural strengthening measures.

The second issue was related to deterioration of the fabric of the external finishing stone, assumed to be only due to natural factors. However, study of many archival photographs from the early twentieth century dated 1909–10 (Mukherjee 1910) indicated that deterioration was not as pronounced as anticipated over six hundred years of the life of the monument. Why, then, were the sculptures in such an advanced state of decay in the following

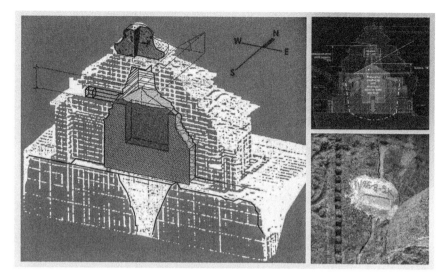

Figure 1.3 Illustrations showing structural condition of the Sun Temple's audience hall interior with the filling of sand and the nature of lateral thrusts on the walls; an image of the breakage in tell-tale markings confirming the presence of the stresses. Source: Author, 2010

hundred years? (see Figure 1.4). A report from ASI's Science Branch indicated that the dominant minerals in the reddish black khondalite sandstone, which forms the external finish of the structure, are alumina and ferrite, both of which are soluble salts (ASI Science Branch 2010, 11). The standard practice of cleaning sculptures aimed to remove salt depositions by the sea breeze is called the paper pulp treatment, which absorbs all soluble salts through capillary action. Being chemically inert, the alumina layer had actually played a role in protecting the sculptures over time. The removal of this layer in every annual cycle of cleaning, along with the foreign salts, had exacerbated the deterioration. The second important action was therefore, ironically, to stop the cleaning of the stone, and measure the surface loss (Sanyal 2012, 214). Monitoring the fabric over three annual cycles showed marked lowering of deterioration rates.

Documenting the state of conservation at Ahom sites

The overall conservation approach for Sivasagar relied on documentation of structures at Sivasagar, which addressed many issues related to inadequate comprehension, interpretation, and presentation of the sites to the public. It also helped to prevent ad hoc interventions and decision-making. This, in turn, marked a shift from hasty interventions that could

Figure 1.4 Loss of surface fabric in sculptural relief at the Sun Temple's upper superstructure due to both natural action as well as paper pulp treatment to extract soluble salts. Source: Author, 2010

potentially be detrimental to the built heritage in the long run, to the creation of a baseline that documented the condition and previous interventions in detail. The documentation and analysis, on the one hand, helped enrich the knowledge and understanding of the sites; on the other, it also helped to ascertain current material and structural condition of these sites before any interventions could be undertaken. This was critical to the project because earthquakes had already impacted the structures. Their exact condition needed to be known and documented for future seismic preparedness, by rendering visible the vulnerable parts of the structures through ichnographic representation.

As discussed earlier, developing in-house expertise in ASI and generating a shared comprehension of the site was a desirable outcome of the process of documentation. Due to scarce resources, basic tools and measuring instruments were used and the exercise of documentation doubled as hands-on training for technical staff. Methodologically, the sites were recorded in their existing state using fixed level triangulation. Each architectural drawing included information of current conditions of various components, and recorded alterations and missing parts of the structures. All interior space was separately documented including information on current condition, in addition to the overall site and structure. This allowed understanding of certain physical issues and deformities that impacted the entire structure and assess damage. Identification of all types of damage also helped in developing a comprehensive legend for condition mapping. These ranged from biological and water damage to mechanical stress, as well as incompatible repairs and interventions.

The documentation revealed the defence system of Ahoms in Rangpur, within which the palace Talatal Ghar and the entertainment pavilion Rang Ghar were located. This enhanced the interpretation of the site as a unified whole instead of two isolated monuments. In Garghaon Palace, several points of structural distress from an early twentieth-century earthquake were recorded. The foremost of these related to the shift of one structural grid at the fourth (top) level. This was, hitherto, unmapped but only revealed through the rigorous condition documentation (Figure 1.5). It informed the team about the risk of visitor access, particularly to the second floor. In Rang Ghar, cracks that were assumed to be surface fissures were revealed to be structural since their profiles continued throughout the section of the roof and across floors. Most significantly, the documentation of the sites at Sivasagar brought the team up-to-date with its latest state of conservation. As a result, all decisions taken thereafter and for the future could be done in a systematic and informed manner. It established a continuous process of

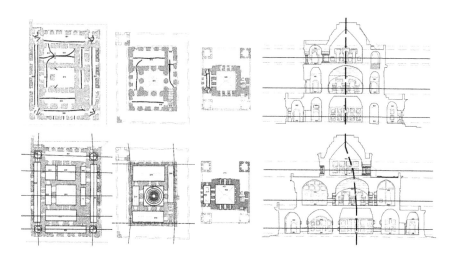

Figure 1.5 Drawings showing existing state documentation of the palace in Garghaon, Sivasagar and an overview of structural assessments; the sections drawn (right) demonstrate how the structural grid in the building shifted after the earthquake. Source: Author, 2012

planning for conservation by recording current state and previous interventions at sites.

Conclusion: continuity and change

The approach for conservation in the cases discussed above can be considered instructive for practice in many ways. The sites themselves are representative of two extremes of ASI's responsibilities – one, a World Heritage Site under public scrutiny, and the other, a relatively unknown site of immense regional significance. The sites also varied in their funding allocations and staffing capacities within the concerned ASI circle offices. However, in both cases, it was observed that a common, subjective way to meaningfully apply academic rigour and research for technical intervention is possible within various enabling or constraining situations. This shows that the ASI, as an organisation, is capable of a shift towards sound professional action in conservation. The cases also illustrate applications of current professional approaches in India, which increasingly work across the social and physical sciences, as well as the humanities, to comprehensively understand sites, their contexts, and components.

Both these instances represent the potential for social responsibility that the state can assume in the field of built heritage, transcending major limitations of outsourced private consultancy. They demonstrate how continuous engagement of a constitutionally mandated professional agency is a necessary requirement for heritage conservation, especially since heritage is considered a social good. Further, these cases illustrate that the ASI is capable of demonstrating competency in conservation approaches for heritage without having to use resources on expensive private consultants. There is greater flexibility when projects do not have to rely on contracts delimited by 'scope' or 'consultancy fee'. Both the Sun Temple and Ahom monuments demonstrate a redefinition, rediscovery, and reorientation of priorities, which are painstaking and protracted, and can primarily be effected through institutional mechanisms, rather than private consultancy firms' 'highest-revenue-in-shortest-time' preoccupations.

Sadly, however, an anti-conservation rhetoric is increasingly gaining momentum, which sees heritage sites as tourism consumables on the one hand, and a hindrance to urban development, on the other – arguing that they occupy land that could potentially be used for remunerative infrastructure development projects. More worrying are official positions intending to relax the legal framework of protecting heritage sites (Sanyal 2017, ASI 2016). As a result, the roles and responsibilities of the ASI, which is one of India's foremost custodians of built heritage, stand to be undermined considerably. Among private practitioners and urban development agencies alike, undermining ASI's professional capacity has been an effective route to possibly securing consultancy work or enticing public stakeholders with promises of 'better practices'. Popular arguments against ASI continue to claim that it lacks competence, intent, and motivation in effectively managing heritage sites. Unfortunately, some independent scholars also endorse such positions against the ASI (Gupta 2017).

In light of the cases discussed above, the presupposed claim on the ASI's 'lack of motivation' seems unsubstantiated, since this work demonstrates that the agency is quite capable of developing new and relevant approaches to conservation. However, it must be appreciated that this can only be made effective through internal institutional change and not by dissolution or diminishing its mandate to privilege private outsourcing. While it is true that the ASI has earned a negative reputation over time due to its resistance to change, I argue that such resistance is not insurmountable. Working within the ASI on these historic sites contradicts many of the perceived misgivings about the agency. Through the processes applied to sites under the ASI's custodianship, I demonstrate that conservation, as management of change, is not a static project – instead, it is a continuous and ongoing process. An organisation like the ASI must be upheld and reinforced with in-house professional capacity. It should retain its role as the forerunner in the responsible management of heritage. Through these examples, as microcosms, we can see that the ASI has tremendous potential to absorb change, as well as develop professionally competent conservation practices. It has the capacity to embrace and expand the current professional practice of heritage conservation, which could mark a significant and positive departure from business as usual.

Acknowledgements

The author wishes to thank all his colleagues at ASI who were instrumental in the support for carrying out these two projects, particularly Dr. D.V. Sharma, Janhwij Sharma, (late) Dr. Subash Khamari, Dr. Ashok Patel, Dr. S.K. Manjul, Dr. G.C. Chauley, Abhijit Deka and Rakesh Kumar. In addition, he is indebted to the invaluable contribution of conservation architects Somi Chatterjee and Saparya Varma in helping materialise and implement the Sivasagar project.

Notes

1 Article 49 in the Directive Principles of State Policy in the Constitution of India pronounces, 'It shall be the obligation of the State to protect every monument or place or object of artistic or historic interest...' This is further reflected in the Seventh Schedule, within both the Union (Article 67) and State Lists (Article 12).
2 The World Heritage sites of the Mountain Railways of India (inscribed 1999) and the architectural work of Le Corbusier in Chandigarh (2016) are not managed by the ASI
3 Deposit works are a form of consultancy and partnership to execute conservation work, where in-house expertise may be absent within the site's custodian

organisation. Recent examples include Ta Prohm temple in Siem Reap, Cambodia, and the National Library, Kolkata (2010-11), among others.

4 K.S. Behera's article critically reviews a traditional text called the *Madala Panji*, which describes the physical characteristics of the Sun Temple. Although the text is claimed to be from the sixteenth century, he observes that it contains descriptions of physical interventions from the twentieth century. This makes its authenticity doubtful.

5 Local officials reported that the plan drawings for the Sivasagar structures were prepared originally in the 1930s–40s; they were retraced in 1964 and 1983, as per the present records (ASI Guwahati Circle 1983).

6 For a detailed assessment of this structure, see Sanyal, Saptarshi (2013), 'Paradigms for Structural Conservation: Observations and Approaches for the Sun Temple, Konarak' in *Context*, Special Issue on World Heritage Sites in India: January 2014 Volume 10 (2), 109–118.

7 Square in plan, the *jagamohana* or audience hall measures thirty metres a side externally, thirty-seven in height from the plinth level of four and half metres. The roof forms a square hollow stepped pyramid in three stages that is thirty-six metres across at its broadest level and eighteen metres high. Its circular crowning element, the *Amlaka*, is thirteen metres in diameter and eight metres high.

8 Feilden's report estimated a pressure of 460 kilonewtons per square metre on its foundations. He states that while this could be considered high for a rapidly constructed modern building, for the Sun Temple, which was built over at least twelve years, the gradual consolidation of the soil, which is granular laterite, would have occurred even as the structure was being built.

9 A tell-tale mark is a linear piece of glass about 60 mm × 25 mm, 4 mm thick, pasted with neat cement on a building's structural crack. If the glass shows a fissure or break across its cross section, this is indicative of mechanical movement or expansion of the structural crack.

References

Allchin, F.R. 1978. "Monument Conservation and Policy in India", *Journal of the Royal Society of Arts* 126 (5268): 746–765.

Anand, Rita. 2015. "Digging its Own Grave", *Down to Earth*. Last modified June 7, 2015. https://www.downtoearth.org.in/coverage/digging-its-own-grave-32700

ASI. 1983. *Papers on Protected Monuments, Sivasagar*. Guwahati: Guwahati Circle.

ASI. 2006. *Ta Prohm Temple: A Conservation Strategy*. New Delhi: Archaeological Survey of India, Office of the Director General.

ASI. 2011. "Alphabetical List of Monuments", *asi.nic.in*. Last modified 2011. https://asi.nic.in/alphabetical-list-of-monuments/

ASI. 2016. "Amendment of the Ancient Monuments and Archaeological Sites and Remains Act, 1958 (24 of 1958). F. No. 1/26/2016-M. Note for the Cabinet", *asi.nic.in*. Last modified 2016. http://www.asi.nic.in/scan0612.pdf

ASI RD-ER. 2011. *Comprehensive Conservation Plan for Ahom Monuments in Sivasagar*. Unpublished Report. Kolkata: ASI Regional Directorate, Eastern Region.

ASI Science Branch. 2010. *Materials Analysis of the Sun Temple, Konarak*. Unpublished Report. Dehradun: Archaeological Survey of India.

Bach, Lee. 2002. "Heuristic Scholar: Heuristic Inquiry and the Heuristic Scholar", In *Counterpoints, Vol. 183, The Mission of the Scholar: Research & Practice: a tribute to Nelson Haggerson*, 91–102. Pieterlen: Peter Lang AG.

Barpujari, H.K. 1990. *The Comprehensive History of Assam*. Guwahati: Publication Borad, Assam.

Behera, K.S. 1996. *Konark: Heritage of Mankind*. New Delhi: Aryan.

Bose, N.K. 1926. *Konarak'er Vibaran*. Translated and Compiled by Prasenjit Dasputa and Soumen Pal. Kolkata: New Age.

Casey, Ethan. "At the Altar of Acrimony", *Outlook*. Last modified December 27, 1995. https://www.outlookindia.com/magazine/story/at-the-altar-of-acrimony/200457

Chauley, G.C. 1997. *Sun Temple of Konark: History of Conservation and Preservation*. New Delhi: Om.

Croci, Giorgio. 1995. *Structural Analysis of the Jagmohana of Surya Deul at Konark*. Paris: UNESCO.

Donaldson, Thomas. 2005. *Konark: Monumental Legacy*. New Delhi: Oxford University Press.

Feilden, B.M and P. Beckmann.1987. *Konarak. Report on the Structural Condition & the Preservation of the Stone*. London: ICCROM.

Fergusson, James. 1860. *Sketchbook and Paintings of the Black Pagoda*. Unpublished Document. Kolkata: Asiatic Society.

Gait, Edward A. 1906. *A History of Assam*. Calcutta: Thacker, Spink and Co.

Gogoi, Shrutashwinee. 2011. *Tai Ahom Religion: A Philosophical Study*. Unpublished Doctoral Dissertation. Guwahati: Gauhati University.

Gupta, Narayani. 2017. "With Custodians Like These", *The Indian Express*, February 22.

ICOMOS. Advisory body evaluation: World Heritage List no. 246. Last modified 2020. https://whc.unesco.org/en/list/246/documents/

Jarrett, H.S. and Jadunath Sarkar. 1949. *Ain-i-Akbari of Abul Fazl-i-Allami Vol II*. Kolkata: Royal Asiatic Society of Bengal.

Malbon, Elizabeth S. 1983. "Structuralism, Hermeneutics and Contextual Meaning", *Journal of the American Academy of Religion* 51 (2): 207–230.

Ministry of Law and Justice. 2015. *The Constitution of India*. New Delhi: Government of India (Legislative Department).

Mitra, Debala. 1968. *Konarak*. New Delhi: Archaeological Survey of India.

Mukherjee, Asutosh. 1910. Album on the Sun Temple, Konarak. Unpublished Photographs. Kolkata: National Library.

Phukan, Jogendra N. 1973. *The Economic History of Assam under the Ahoms*. Unpublished Doctoral Dissertation. Guwahati: Gauhati University.

Sanyal, Saptarshi. 2012. "Exclusion and Efficiency in Measuring Heritage Conservation Performance", In 6th International Seminar on Urban Conservation: *Measuring Heritage Conservation Performance*, 212–221. Rome: ICCROM.

Sanyal, Saptarshi. 2017. "Disrespecting Heritage", *The Indian Express*, January 25.

Sharma, D.V. and Saptarshi Sanyal. 2010. *The Sun Temple Konarak: Past, Present and Future*. Bhubaneswar: Archaeological Survey of India.

Starza-Majewski, O.M. 1971. "King Narsimha I before his Spiritual Preceptor", *Journal of the Royal Asiatic Society of Great Britain and Ireland* 2 (1): 134–138.

Swarup, Bishan. 1980. *Konarka: The Black Pagoda of Orissa*. Delhi: Ramanand Vidya Bhawan.

Thakur, Nalini. 2007. "Hampi World Heritage Site: Monuments, Site or Cultural Landscape", *Journal of Landscape Architecture* 5 (1): 32–37.

WHC. 2018. "World Heritage List: Sun Temple, Konarak", *whc.unesco.org*. Last modified 2018. http://whc.unesco.org/en/list/246

WHC. 2020. "Properties Inscribed on the World Heritage List in India", *whc. unesco.org*. Last modified 2020. http://whc.unesco.org/en/statesparties/IN

2 The role of Indian National Trust for Art and Cultural Heritage in heritage conservation in India

Divay Gupta

Introduction

India has an abundance of rich history, culture, traditions, and heritage, but a mere 0.7% of its total built heritage is under some sort of official protection (Khosla, Saini, et al., 2017). The central government's Archaeological Survey of India (ASI), its state counterparts, and the local urban or rural administrative bodies (ULBs), are all custodians of about ten thousand monuments and archaeological sites combined. This leaves out a vast majority of built heritage, conservatively estimated at close to one million sites, from any means of protection or proper conservation. Historic cities or precincts have not fared well either; out of the eight thousand cities and towns in India, barely twenty are recognized as heritage cities and even fewer have any special planning considerations within their master plans. Out of this plethora of 'unprotected heritage', many sites are often in use and represent the 'living heritage' of the country. Yet they remain unprotected and vulnerable to insensitive modernization, urbanization, and unchecked development. On the other hand, the majority of India's recognized architectural heritage and sites, even though legally protected, in effect remain unprotected in the absence of strong enforcement mechanisms (Chainani 2007).

In such a scenario, the work of non-governmental and advocacy groups becomes very important. This chapter focuses on the Architectural Heritage Division (AHD) of the Indian National Trust for Art and Cultural Heritage (INTACH). It traces the evolution of built heritage conservation in INTACH over the last twenty years and how that has shaped conservation philosophy and practice in India more generally. Additionally, this chapter argues how INTACH, through its built heritage conservation programme has been trying to promote an India-centric approach to conservation, which draws from international approaches, but takes into consideration indigenous ways of conservation.

Background: INTACH

Evolution

Though INTACH was only established in 1984, it owes its existence to a colonial entrepreneur by the name of Charles Wallace, who at the time of his death in 1916 left a bequest to the British treasury for the promotion and study of art and culture within British India (Menon 2009). Wallace's relatives contested the will and the case dragged on in the courts for over fifty years, when in 1967, the High Court in London ruled that the bequeathed funds should be paid to the successors of the Treasury of British India, which by then were three separate nations: India, Pakistan, Burma (and later in 1971 Bangladesh would also be added to this list). Of course, India got the lion's share of the appropriations based on population. In 1968, a memorandum of understanding was signed between the then Indian Finance Minister Morarji Desai and the British Home Secretary to transfer the amount of roughly nine million British pounds, to be utilized in accordance with the original intent of Wallace's will. However, the funds languished in the London branch of the State Bank of India until 1982. With the efforts of the then Prime Minister, Indira Gandhi, and Pupul Jayakar (cultural activist and writer), those funds were ultimately brought to India and used toward setting up INTACH, which would be recognized and supported by the Government of India but function largely as an independent agency modelled after the National Trust in the United Kingdom. Thus, INTACH came into being on 27 January 1984, with Rajiv Gandhi as its first chairman, who at that time was the General Secretary of Indian National Congress and later became the Prime Minister in October 1984 (Menon 2009).

Mission

The vision of INTACH's founding members was to create a membership organization to stimulate and spearhead heritage awareness and conservation in India. In the last three decades, INTACH has pioneered the conservation and protection of India's natural and cultural heritage and is today the largest membership organization in the country dedicated to heritage conservation.

The mission statement of INTACH covered an ambitious vision, from protection of India's natural heritage to conservation of art and architectural heritage. INTACH's mission is based on the belief that living in harmony with heritage enhances the quality of life, and it is the duty of every citizen as laid down in the Constitution of India (Ranjan 2009). The organizational structure demanded two distinct functioning divisions: one related to advocacy and awareness through its members, and the other related to professionalism through its technical divisions which would focus on the technical aspects of conservation of a range of heritage type. Thus came into existence the following divisions: architectural, natural, art and material heritage, and later, intangible, heritage education, heritage tourism, and craft and community divisions. Today there are a total of nine divisions focusing on different aspects of heritage conservation.

INTACH today

Within a short span of a few decades, INTACH has emerged as the largest heritage-based non-governmental organization in India and perhaps in the world, with more than 200 chapters spread across India, 9 regional conservation centres, 10,000 volunteer members, and employing more than 150 conservation professionals. The agency has also established its own charter for conservation, based on its body of work across architectural, material, and natural heritage conservation. The charter attempts to redefine 'heritage' to include traditional built typologies in various forms including heritage zones, precincts, bazaars, civic buildings, gardens and townscapes – the majority of what constitutes 'unprotected' heritage in the country. The charter recognizes that conservation in India is derived from not only Western conservation theories and principles introduced through colonialism, and guidelines formulated by global agencies such as UNESCO and ICOMOS, but also indigenous knowledge systems and building traditions (Menon 2009). These indigenous practices vary regionally and cannot be considered as a single system operating within all of India. This necessitates viewing conservation practices in India as a multicultural activity that may be region-specific.

The Western ideology of minimal intervention underpins the official and legal conservation practice in India and is appropriate for conserving protected monuments (Menon 2009). However, conserving vernacular built environments provides the opportunity to use indigenous practices and maintain their contemporary relevance. Before undertaking conservation, therefore, it is necessary to understand what conservation practices are indigenous to a particular region and how they can be applied to sites in that region.

Among the tasks undertaken by INTACH are restoration of monuments and their management, advocacy for heritage property conservation, public awareness through educational and outreach programs, establishment of heritage clubs in schools, and identification and listing of unprotected sites. Particularly, the agency works on historic sites that fall outside the purview of the ASI and other government agencies. Often, local authorities give responsibility for the upkeep and restoration of historic properties to INTACH directly, as they are custodians of a number of heritage assets but do not have in-house expertise in conservation. INTACH has tried to fill this gap in expertise towards conservation of 'unprotected heritage', particularly sites that fall under the purview of local authorities and state departments. Over the years INTACH has been able to conserve or intervene in more than one hundred such sites.

Role in professional practice

One of the major contributions of INTACH has been towards creating a cadre of conservation professionals. Heritage conservation was not an established discipline of study in higher education systems in India until the 1980s. This began to change in 1984 when INTACH, through the British Council and the Charles Wallace Trust, helped facilitate scholarship for Indians to study conservation in the United Kingdom (Menon 2009). Later, INTACH collaborated with the School of Planning and Architecture (SPA, established 1959) in New Delhi and helped set up the country's first postgraduate program in Architectural Conservation. The professionals that graduated from these Indian and British programmes slowly grew in number and brought conservation practice into the mainstream. The professional and technical expertise they provided began to transform conservation practice in India. Many of these professionals engaged with a range of manifestations of heritage: built, natural, material, and intangible, and brought about integration within the diverse aspects of heritage conservation and management. The conservation educators and professionals who were closely associated with INTACH included AGK Menon, KT Ravindran, Romi Khosla, Nalini Thakur, Rajat Ray, Anuradha Chaturvedi, and Priyaleen Singh, among others. Although small in number, these educators and practitioners helped define INTACH's role while confronting bureaucratic hurdles, and all the while streamlining conservation practice.

As demand for conservation escalated, INTACH established an in-house cell of conservation architects in 1999 under its Architectural Heritage Division (AHD). Until then, all the projects had been outsourced to external experts and consultants. The setting up of the AHD was not acceptable to many conservation architects as external consultants and senior experts had benefitted from the earlier arrangement, while others were concerned about the growing presence and influence of INTACH in professional practice of conservation. Some started to see the agency as a professional competitor and an impediment to their professional growth and commercial interests. Many critics also mourned the loss of INTACH's initial mission and voluntary zeal. In time, the Architectural Heritage Division began to be seen largely as facilitator and knowledge partner (and not a competitor) by most conservation professionals, and today garners much support from practitioners. More recently, in 2016, as a way to standardize practice, the Architectural Heritage Division developed a standard schedule of rates for conservation works, which is available as a reference manual for all conservation professionals from INTACH's official website (www.intach.org). Today INTACH is the largest organization to employ and use the services of conservation architects, many of whom it helped train.

The Architectural Heritage Division (AHD)

The Architectural Heritage Division (AHD) was among the first divisions created by INTACH in 1984. Its mandate was clear: to identify and conserve unprotected heritage. The first mandate, identifying unprotected built heritage, was more daunting than conservation, as there was no baseline data available. The prime focus in the initial years of the AHD was related to identification or 'listing' of undocumented historic buildings and sites. Though a few isolated conservation works were undertaken during this time, they were met with limited success. Some examples include the Bhuli Bhatari Palace complex in Delhi, Raja Ghat in Varanasi, and Raj Mahal in Chanderi. Many of these initial conservation efforts focused mainly on restoration without integrating (or fully implementing) any long term plans for their reuse, operation, and maintenance within the conservation plan, due mainly to lack of funds or administrative capacity of sponsoring government departments. Based on this experience with conservation projects, the AHD focused its efforts on documenting historic buildings, much of which was conducted from 1984 to 1999 by student volunteers or heritage enthusiasts. Today, the Architectural Heritage Division has an inventory of more than seventy-thousand sites in five-hundred-and-fifty-two cities and towns across twenty-six states. This task, however, is far from complete, and the quality of listing varies across different contexts: they often lack rigour, and in many cases are not comprehensive. The initial listing of Odisa, for example, mainly covered the temples, and omitted other typologies, while listing of the Shekhawati region in Rajasthan mainly documented painted *havelis*.

Despite these shortcomings, the listing project generated a record of historic properties that never existed before, and established conservation priorities and benchmarks in recording built heritage across the country. More importantly, the listing process redefined built heritage as not exclusive to just monuments, but expanded the scope to include a variety of vernacular heritage types including bazaars, warehouses, army barracks, and residential dwellings, among others. Unfortunately the bulk of the inventories have remained just that – they have not been notified by local administrative units despite pressures from the local INTACH chapters. Except for some success stories from Mumbai, Pune, Nagpur, Delhi, Gorakhpur, and Cochin, much of the rest of the unprotected heritage remains vulnerable to unchecked forces of development (Khosla, Saini, et al., 2017).

Besides individual listings, the AHD has also identified 'heritage zones', defined as areas of special architectural, historical, or cultural significance which needed to be conserved. By 1990, fifty such zones were studied and documented, including those in Varanasi, Chanderi, Leh, Cochin, Mathura, Ujjain, and Bhubaneswar. Though at that time the concept of heritage zones was not popular, it is now being recognized and its integration with developmental or master plans is taking shape in cities such as Delhi, Cochin, Varanasi, Cuttack, and Jaipur. Both the listing process and heritage zone documentation were based primarily on existing models from the UK, France, Australia, and the USA. This exposed the emerging field of heritage conservation in India to additional important ideas of environmental sustainability, cultural pluralism, and diversity. Since then, these concepts have become the bedrock of INTACH's conservation philosophy.

Once the identification and documentation phase was complete, the AHD began to shift focus to its second mandate in 2000: conservation and restoration of 'unprotected built heritage'. By the late 1990s, INTACH had already established itself as an expert in heritage policy and documentation, but was being criticized for its limited experience in actually executing conservation plans. The turning point came with the 2001 Gujarat earthquake. The devastating earthquake in Bhuj not only created havoc in terms of loss of life and property, but also damaged much of the historic properties in the region. Consequently, the AHD initiated an assessment of the earthquake's damage on historic buildings, conducted surveys, and developed plans for the restoration and rehabilitation of damaged buildings, especially in Bhuj and Morvi. Based on this survey report, INTACH convinced the Gujarat government to introduce a financial compensation package for historic property owners to enable repairs and restoration work. This was the first time that such a post-disaster initiative was announced by any state government, and is considered a significant achievement by INTACH. Many of these earthquake buildings were restored by the AHD at the request of various governmental bodies and private owners. From here on most of the focus of the AHD shifted to its conservation mandate. With the economic liberalization of India in the 1990s, funds had also finally started to be allocated for actual conservation work by central and state governments. Consequently, several government and public works departments including the post office, railways, and army began approaching the AHD to restore their heritage buildings for which these departments lacked expertise.

The INTACH Charter

The conservation phase of the AHD led to the creation of many cultural assets and direct experience with conservation practice while testing Western conservation principles and norms in the Indian context. These were, at that time, largely based on ICOMOS Charters, especially the Venice Charter, which was also the basis for *Guidelines for Conservation*, developed by Sir Bernard Feilden in 1989 as a technical manual for INTACH (Fielden 1989). This was an updated version of John Marshall's manual for the ASI from 1924. The difference between the two manuals was minimal, except that Sir Feilden introduced the concepts of urban conservation and reuse of historic structures, while Marshall's manual was limited to archaeology (Marshall 1923). Through its years of conservation practice, the AHD has been deliberating on an appropriate conservation philosophy that is rooted in the varying Indian contexts. With the experience gained from the post-earthquake conservation work in Gujarat, and from a series of other restoration projects, INTACH decided to reconcile the worldview in conservation within the Indian context, to make it more consistent with international approaches related to conservation while retaining and enhancing the indigenous conservation practices. The result was the INTACH Charter for 'unprotected heritage', which was adopted in 2004 by INTACH's general assembly during its commemoration of twenty years.

Though the basic conservation approach was aligned with international conservation practices, the INTACH Charter is unique in two distinct aspects: first a de-emphasis on the need to retain patina, and second, its rejection of minimal intervention by allowing the residents/craftspersons to restore and rebuild using traditional knowledge systems without necessarily having to re-create the original historic fabric. So while the Marshall manual in some instances permitted reconstruction of decorative elements or missing architectural features in Islamic buildings (given their geometric simplicity), it debarred any sort of replication of figurative carvings in temples. The INTACH Charter argues for the latter on the basis of *shilpa shastra*, the ancient manuals of temple construction and iconography.

Using this approach, the AHD successfully restored the fire gutted British Residency building in Srinagar, which was reconstructed fully, mainly from local memory. The classical restoration was not possible due to limited available documentation, including photographs. After the building envelope was reconstructed incorporating remaining burnt walls of the historic building, the local artisans were encouraged to reconstruct the building from collective memory using traditional building crafts of *pinjrakari* (lattice work), *khath-ambandi*, (composite wooden ceiling), and *naqashi* (wood carving). This they achieved rather successfully as the building they created closely matched the original after some historic photographs were discovered by chance in a local archive two years after the restoration was completed (Figure 2.1).

In conserving the palace complex in Chanderi (Central India), an entirely new section of the palace was reconstructed using local building materials, and salvaged historic materials: stone columns, beams, and brackets obtained from incongruous more recent constructions on the site (Architectural Heritage Division INTACH 2019). The same materials were also used in creating new colonnades that both served as structural buttresses and corridors connecting different blocks in the complex.

Some traditional conservationists have been critical of this type of reconstruction based on conjecture (both within and outside of INTACH) as having 'lost our ways' or deviating from established conservation norms, but as John Stubbs has acknowledged:

> The field of architectural conservation practice, rich as it is now, can only be enhanced by a fuller appreciation of how the so-called East-West philosophical debates of heritage protection translate into the present array of potential solutions for both conserving and interpreting cultural heritage sites.
>
> (Stubbs 2009, 270)

INTACH, through the work of the AHD, has contributed substantially to that discourse (Figure 2.2).

British Residency after fire in 2000

Restored British Residency in 2004

British Residency before fire in 1994

Figure 2.1 Restoration of the British Residency at Kashmir Emporium, Srinagar.
Source: INTACH J&K, 2000–2004

The AHD today

Today, the AHD is considered to be a major conservation stakeholder along-side the ASI. It is bringing greater order and accountability to its work, and is among the few organizations that encourage peer review of all projects by external experts. Much of the passion of the early years still remains, but today its work is supported by multiple professional disciplines and technology, including instrumentation and structural diagnosis to establish significance or condition of the historical buildings. In the recent years, the AHD has also recognized that documentation, listing, or conservation is not a one-time intervention, but a continual process. This is evident from the current state of many buildings that were restored by INTACH in the past, which have returned to their pre-restoration state in the subsequent years. It is easier to get capital grants for restoration but more difficult to get any recurring funds for their long term upkeep and management. Thus, INTACH is exploring models and suitable financial instruments to make heritage conservation attractive and sustainable in the long term. Perhaps many of the solutions for heritage conservation lie not in the conservation labs or architectural drafting boards, but with economists, anthropologists,

Figure 2.2 Before and after images of the restored Raja Mahal at Chanderi. Source: INTACH, 2006–2008

or sociologists. INTACH has now started a dialogue with professionals in those fields to make conservation more economically viable and socially relevant. Additionally, with the amended AMASR Act of 2010, the role of INTACH, and especially that of the AHD, has been redefined in the context of monuments protected by the ASI. Special bylaws were framed by INTACH on the basis of the parameters established by the new legislation. These bylaws were specific to the regulated area of the ASI monument. These were unlike the city level bylaws of the urban local bodies, and controlled the new construction in the area in terms of facades, usage of materials, colours, height, excavations, and restrictions on underground construction.

Though the AHD continues its work on listing, documentation, conservation, and stakeholder engagement in its projects, it has gone through three distinct stages in its evolution over the last 30 years with each stage lasting about a decade. As noted earlier, the first stage focussed mostly on listing, documentation, and preparation of conservation plans. The listing particularly expanded the scope of heritage types and clarified the distribution, typologies, and state of repair of historic sites. The second phase at the turn of the twentieth century focused mainly on creating cultural assets through their actual conservation, restoration, and reuse. This phase provided the AHD with an opportunity to carry out conservation work and test the various approaches and develop its own charter addressing the unique context of India and its continuing living and craft traditions. The third stage is currently active, at a time when INTACH is engaging more with policy intervention and management of sites by looking at the state of the built heritage in India and addressing the risks and challenges it is facing today in a fast urbanizing and globalizing world. INTACH is also looking at institutionalizing heritage and integrating it with urban planning to explore financial mechanisms for investment in historic properties. Another important agenda of the AHD is to involve local communities in the management, conservation, and upkeep of their heritage. After laying the groundwork for architectural conservation in India, the Architectural Heritage Division of INTACH now intends to look at how it can strengthen these foundations by facilitating conservation processes and promoting good practices.

Reflection

INTACH, and specifically the Architectural Heritage Division, have been an integral part of architectural conservation in India since its inception.

> In the course of the last 25 years, the single achievement of INTACH has been in addressing this agenda, it has transformed the conservation scene in India both as a technical discipline and as a public movement. This is visible in the manner in which our contemporary society deals with its past and the meaning it assigns to heritage in planning for its future growth and development.
>
> (Menon 2009, 6)

Today INTACH is creating an ever-expanding countrywide network of conservation professionals for providing technical assistance to heritage property owners: advice on emergency repairs, restoration, adaptive reuses, preparation of conservation plans, and available financial options for those seeking assistance for conservation work.

An important lesson learnt by INTACH in all its projects is that conservation is a long-term and at times a lifetime commitment – even for a specific building, it goes beyond the conservation plan or even the implementation stage, and thus the resource planning has to be done accordingly. It has only recently been realized that financial feasibility, operation, and maintenance in conservation plans are equally important as are the technical specifications of brick and mortar. Another realization has been that conservation is as much a technocratic process as it is a professional exercise. Thus a conservation architect also needs to play the role of a negotiator between political context, decision-makers, and conservation norms.

Author's note

I would like to place on record that I have worked for INTACH in various capacities since 1995. This association has given me a unique vantage on the functioning of the agency. Despite that, I have tried to be non-partisan and presented an account of the agency without any biases, to the best of my abilities. Since there is no official history of INTACH, this chapter is mainly based on my personal experience working with INTACH for over 25 years. The main source for INTACH's establishment is from a booklet by Professor AGK Menon on Indian National Trust for Art and Cultural Heritage in 2009 as part of INTACH's silver jubilee celebration. Most of the other information is from INTACH's recently published works such as *A Momentous Journey, 30 Years of Conserving Architectural Heritage* and the unpublished report *State of Built Heritage, Case of the Unprotected*, both prepared under the supervision of the author. The views expressed here are those of the author.

References

Architectural Heritage Division INTACH. 2014. *A Momentous Journey, 30 Years of Conserving Architectural Heritage*. New Delhi: INTACH Publications.

Chainani, Shyam. 2007. Heritage & Environment : An Indian Diary. Mumbai: Urban Design Research Institute.

Fielden, Bernard. 1989. *Guidelines for Conservation: A Technical Manual*. New Delhi: INTACH Publications.

Khosla, Rekha, Malvika B. Saini, et al. 2017. *State of Built Heritage, Case of the Unprotected (SoBHI)*. New Delhi: INTACH Publications.

Marshall, John Hubert. 1923. *Annual Report of the Archaeological Survey of India*. Calcutta: Government of India, Central Publication Branch.

Menon, A.G. Krishna. 2009. *Indian National Trust for Art & Cultural Heritage 1984–2009*. New Delhi: INTACH Publications.

Ranjan, Neena, ed. 2009. *INTACH Journal of Heritage Studies*. New Delhi: INTACH Publications.

Stubbs, John H. 2009. *Time Honored: A Global View of Architectural Conservation*. Hoboken, New Jersey: John Wiley & Sons, Inc.

3 Heritage management and conservation planning for historic cities in India

The case of Jaipur and Ajmer

Shikha Jain

Introduction

Conscious approaches to urban conservation and heritage management in India are recent. Some local municipal laws in cities in Rajasthan have required the conservation of street-facing facades since the 1970s (The Municipal Council, Jaipur (Building) Bylaws, 1970). Mumbai and Hyderabad are recognized since the 1990s as providing model heritage bylaws[1] for historic Indian cities. However, the preparation of a city-level heritage management plan and recognition of urban conservation planning by the Ministry of Housing and Urban Affairs (MoHUA) (formerly Ministry of Urban Development) in India came much later. The Jaipur Heritage Management Plan prepared in 2007 is recorded as one of the first initiatives of a city-level heritage plan in India followed by similar attempts in Madurai and Varanasi. These singular city initiatives were recognised by the MoHUA and resulted in the inclusion of heritage management plans in the revised city development plan toolkit under the Jawaharlal Nehru National Urban Renewal Mission (JNNURM) in 2013 (Revised Toolkit for Preparation of City Development Plan 2009).

A review of past urban renewal schemes introduced from 2009–2014, and the need for heritage-focused development for historic Indian cities led to the introduction of the National Heritage City Development and Augmentation Yojana (HRIDAY)[2] launched on January 21, 2015 with the aim of bringing together urban planning, economic growth, and heritage conservation in an inclusive manner to preserve the heritage character of 12 identified heritage cities (Guidelines for HRIDAY: Heritage City Development and Augmentation Yojana 2015). This chapter will trace the history of urban heritage planning in India through the case studies of Jaipur and Ajmer (under HRIDAY) since 2015 and examine their urban frameworks and governance, including the challenges faced in contemporary conservation of the city's public and residential spaces. It will elaborate on the heritage management initiatives at the city level that simultaneously influence policy interventions of the Government of India as well as impact phased implementation of large-scale urban conservation projects in these cities. The chapter further informs that sustainability of urban conservation projects resides in anchoring them fundamentally to a city vision plan, along with an inclusive participatory approach in conservation projects.

Historic cities in India and Rajasthan

Historic cities in India show a wide variation in settlement typology (and use of materials) as per geography, climate, and culture of different regions. The northwestern state of Rajasthan in India has several walled cities from the medieval period that have sustained development changes, yet retained their historic character to a large extent. Each sub-region in the state has managed to retain its local character and urban vocabulary shaped by available materials, craftsmanship, and cultural traditions. This notion of preserving heritage in Rajasthan largely bears roots in the royal patronage and continuity of craftsmanship that survived in various forms through the colonial and post-colonial periods. The walled cities of Rajasthan showcase their living heritage worldwide and are exemplars for indigenous modes of preservation that integrate traditional values and practices. The cities and towns of Rajasthan present interesting cases of urban conservation initiatives by the public and private sectors, providing opportunities for income generation by the local craftspeople and residents alike. In such situations, where stakeholders have centuries of association with the sites, it is essential for the conservation professionals to look beyond conventional western guidelines to engage local situations critically to determine effectiveness of intervention strategies.

The author's decades-long experience of working on varied conservation projects with local, state, national, and international bodies[3] is reflected in the following discussion on the urban historic cores of Jaipur and Ajmer. The author prescribes a conservation approach for urban heritage that promotes holistic vision, integrating heritage assets with mainstream planning and development while ensuring inclusiveness.

Jaipur: a heritage management plan for a uniquely planned city

Jaipur, being the capital of Rajasthan, is the focus of the socioeconomic and political life of the state. The city grew rapidly in the decade of 2000, both in terms of physical expansion as well as population that grew from 2.3 million to 3 million. The population of the metro region of Jaipur is expected to reach 4.2 million by the year 2025.

As an exemplar of urban planning, the city is unique and futuristic in its urban form that caters to the idea of Jaipur as a centre of trade and commerce, a creative hub for the arts and crafts and a city with a distinct identity. The historic walled city was placed on the UNESCO Tentative List in 2015, and inscribed as a World Heritage Site in July 2019 at the World Heritage Committee meeting at Baku, Azerbaijan. According to the Statement of Outstanding Universal Value (OUV) adopted by UNESCO upon its inscription, "Jaipur is an expression of the astronomical skills, living traditions, unique urban form and exemplary foresighted innovative city planning of an 18th century city from India". (UNESCO 2019).

The walled city of Jaipur was planned in the year 1727 CE by the ruler Sawai Jai Singh II with a vision for his new capital to be a strong political statement on par with Mughal cities, and to evolve as a thriving trade and commerce hub for the region (Jain 2005, 90). Its evolution can be broadly classified into three phases. The first phase, from the 18th century onward, from the time of Sawai Jai Singh II, is discernible through architectural typologies characteristic of this period such as *havelis*, *haveli-* temples, and *shikhara*-temples. The second phase is marked by 19th-century changes under Sawai Ram Singh II. It was also the time when the colour of Jaipur bazaars changed from the earlier lemon lime wash to a reddish sandstone wash that gave Jaipur its title of "Pink City". The last distinct phase during the early 20th century is marked by the city's expansion beyond the walled enclosure under the reign of Mirza Ismail, the then prime minister to the Jaipur rulers.

After India's independence in 1947, Jaipur became the capital of Rajasthan and around this time the royal residences moved out from the City Palace to Ram Bagh Palace on the southern outskirts. As a state capital, the city further strengthened its economy via trade and tourism, expanding its tourism infrastructure to better accommodate the increasing flux of national and international visitors to Jaipur (Jain, 2010a). The city's tourism economy is primarily dependent on its cultural heritage and local craft industries. Rajasthan Tourism Policy is observed as one of the first in India that includes tax incentives for retrofitting historic properties. This encouraged a number of historic property owners to adapt their *havelis* into hotels or guesthouses without impacting their historic integrity.

Existing governance mechanisms and Jaipur heritage management plan

Today Jaipur Municipal Corporation (JMC) is the main body that manages the historic walled city area of Jaipur and Amber – the old capital towns dating from the 18th century and 16th century respectively. The Archaeological Survey of India (ASI) and State Department of Archaeology and Museums (SDAM) protect all major monuments and sites in and around Jaipur. ASI has 8 protected monuments in Jaipur city while SDAM has a total of 65 protected monuments. Additionally, a special body, Amber Development and Management Authority (ADMA) was created by the Government of Rajasthan in 2005 for the upkeep and maintenance of monuments in Amber that are protected by SDAM. The ADMA is a good example of a government-created agency with a revenue generation model for upkeep of historic sites. The mandate of this body was further extended later to prepare the management plans for the World Heritage

Sites of Jantar Mantar and Amber Fort, and to implement them in a phased manner. A number of non-government bodies such as INTACH (Indian National Trust for Art and Cultural Heritage), including its local chapter in Jaipur, and private organizations like the Development and Research Organisation for Nature, Arts and Heritage (DRONAH), Jaipur Virasat Foundation (JVF), and Indian Heritage Cities Network (IHCN), among others are actively involved in overseeing implementation of conservation works and raising awareness for heritage conservation in Jaipur. They have successfully brought together different stakeholders and partners to work with Jaipur Municipal Corporation (JMC) while coordinating efforts with SDAM and the Department of Tourism.

The Built Heritage Management Plan was prepared for Jaipur in 2007 by the Jaipur Heritage Committee (JHERICO), that was specially created for defining a vision for Jaipur's urban heritage. The plan was commissioned by the Jaipur Virasat Foundation in association with DRONAH. It has been recognized as one of the first initiatives of its kind: a city-level heritage plan in India. These singular city initiatives were recognized by the Ministry of Housing and Urban Affairs that recommended inclusion of city-level heritage plans in the revised City Development Toolkit under urban renewal mission in 2013. This revision in the toolkit was carried out by UNESCO on behalf of the Ministry of Housing and Urban Affairs, and it defined the terms of reference for a Heritage Management Plan to be prepared for any city in India (Revised Toolkit for Preparation of City Development Plan, 2009).

Since 2007, the vision and the phased implementation outlined in the Built Heritage Management Plan for Jaipur has been pursued in various forms through the changing government schemes, detailed project reports, and changing bureaucracies. The Plan listed 1096 heritage structures in 2007, which was updated to around 1500 structures in 2019 during the nomination of Jaipur City for World Heritage. A detailed inventorying is now being carried out by INTACH Jaipur Chapter since February 2020 to document all 1500 structures. Another significant contribution of this Built Heritage Management Plan is its inclusion and absorption in later official plans, such as the Jaipur Master Plan 2025 prepared in 2012 or the Smart City Plan of Jaipur proposed under the Smart City Scheme of MoHUA in 2015. The Built Heritage Management Plan of Jaipur has been recognized as a best practice example by the National Institute of Urban Affairs, Ministry of Housing and Urban Affairs, Government of India in 2015.

Culture-based regeneration strategies

The city's heritage has benefited tremendously from urban regeneration schemes of the Ministry of Housing and Urban Affairs (MoHUA) that involved implementation of urban conservation works for three main bazaars i.e. Chaura Rasta, Tripoliya, and Johri in 2013–14. Conservation of these bazaars along with urban conservation works in the Ghat ki Guni heritage street on the outskirts of Jaipur received a national level award from the Housing and Urban Development Corporation Limited (HUDCO) in 2014.

As per Jaipur Master Plan 2025, the walled city area is a specially designated heritage zone and any conservation work within the walled city is guided by heritage management plans and project reports implementation by government agencies. A heritage cell was created by the Local Self Government Department, Rajasthan[4], which has been revising and strengthening these bylaws for the listed heritage zone, precincts, and structures. There are existing façade control guidelines by Jaipur Municipal Corporation for the main bazaars in the walled city and Chowkri Modikhana[5], one of the historic areas within the walled city. These guidelines are an extension of the municipal council, Jaipur (Building) Bylaws, 1970 (Part V, No. 26) and aid in maintaining the original form and the overall urban character of the walled city. The World Heritage Site of Jantar Mantar, including major urban monuments of Jaipur like Hawa Mahal, City Palace, Jaleb Chowk, and the Town Hall in the buffer area (see Figures 3.1 and 3.2) are protected and managed through their site management plans, which are also a part of the Jaipur City Master Plan 2025. All these aspects have helped Jaipur walled city to retain its authentic historic character and feel. Some

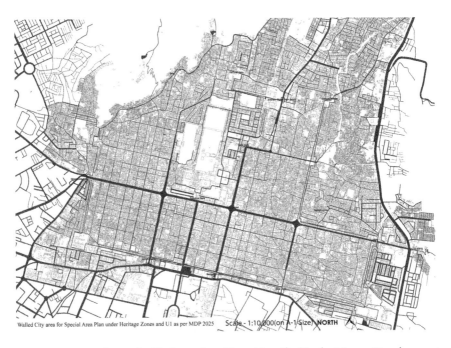

Walled City area for Special Area Plan under Heritage Zones and U1 as per MDP 2025   Scale - 1:10,000(on A-1 Size)   NORTH

Figure 3.1 Map shows the Heritage Area Zone identified in the Master Development Plan 2025. Source: Author

Figure 3.2 View of monuments of Jantar Mantar and City Palace in Jaipur Walled
City from Hawa Mahal. Source: Author

initiatives for conservation and urban renewal in Jaipur since 1971 are outlined in Table 3.1.

The launch of the heritage walk of Chowkri Modikhana was identified
and initiated by INTACH and the JVF in 2001. The JVF has been conducting the walks since its inception and has been bringing visitors to the historic
core of the city. The walks became a milestone which in turn initiated many
urban heritage conservation efforts in later years. Through fundraising initiatives of the JVF, Sankri Gali, a street featured on the walk, was conserved
through the Prince Charles Fund. The building façades and public areas on
the street were restored using traditional methods of lime plaster and stone
works. The local community extended full support to the project and took
over the responsibility to maintain the area. While the JVF continues to
inform the locals in this area and has conducted regular walks for awareness, it took a lot of time for the Jaipur Municipal Corporation (JMC) to
provide collaborative support in upgrading the services and infrastructure
for this walk area. With time, the heritage walk became very popular, as
shopkeepers could now attract foreign tourists, who earlier never visited the
inner areas. The services and infrastructure, and signage and lighting in this
walk area were finally implemented by the JMC in 2013–15 under the last
segment of urban renewal scheme for Jaipur.

Table 3.1 Conservation and urban renewal initiatives in Jaipur

S.No	Year	Organization responsible	Walled City Renewal Initiative
1	1971	JDA (Jaipur Development Authority)	Municipality Act including control guidelines for Urban character of the Walled City
2	1985	Ford Foundation and JDA	Study of Heritage Buildings within walled city
3	1995	Avas Vikas Sansthan and Department of Tourism	Conservation and restoration works of heritage structures
4	1998	JDA	Master Plan 2011 addressing needs of the walled city
5	2001	JMC (Jaipur Municipal Corporation)	Operation Pink, removal of encroachments in main commercial streets of the walled city
6	2001	INTACH and JVF (Jaipur Virasat Foundation)	Heritage Walk in Chowkri Modikhana
7	2001	ADB and JMC	The Asian Development Bank project of infrastructure – reuse of wells and repair work in the walled city/ installation of sewage pipes
8	2002	JMC, JDA and CTP (Country and Town Planning)	Multi-storeyed parking options within the walled city
9	2003	JVF	Jaipur Heritage International Festival
10	2004	Asia Urbs	A revitalization proposal for Chowkri Modikhana
11	Since 2005	Government of Rajasthan	Conservation of City gates, Amber Palace, Jaleb Chowk in City Palace and Ghat Ki Guni heritage zones, lighting of several monuments, making of Heritage Acts and Laws and Empanelment of Conservation Architects to prepare conservation proposals for Grade I and II monuments.
12	2006–2007	JHERICO, JMC	Heritage Management Plan prepared with 1096 listed structures and Conservation project of Amber Fort and Hawa Mahal
13	2007–2008	ADMA and Dept. of Archaeology & Museum, Rajasthan	Conservation project of Jaleb Chowk, Jantar Mantar and Ghat ki Guni
14	2009–2010	ADMA and Dept. of Archaeology & Museum, Rajasthan	Architectural Control Guidelines (initiated) and inscription of Jantar Mantar as World Heritage Site

(Continued)

Table 3.1 (Continued)

S.No	Year	Organization responsible	Walled City Renewal Initiative
15	2011–2013	JDA, DLB (Department of Local Self Government), JMC ADMA, JMC	Heritage Management Plan included in Jaipur Master Plan 2025 and Documentation of Crafts as part of UNESCO dossier Conservation of Jaipur Bazaars, Ghat ki Guni and Jantar Mantar Buffer Zone
16	2014–2016 onwards	JMC and Jaipur Smart City Ltd. ADMA, Dept. of Archaeology & Museum, Rajasthan Dept. of Archaeology & Museum, Rajasthan and JMC JMC	Jaipur SMART City Plan includes Walled city, Heritage Management Plan (Best Practice 2015) and bazaars conservation Conservation projects of Heritage Walk, Jantar Mantar Interpretation Centre Jaipur city on Tentative List and Jaipur as UNESCO Creative City Jaipur designated as member of UNESCO Creative City network
17	2017–2018	JMC and Jaipur Smart City Ltd., ADMA	Conservation and Lighting Of bazzars under the Smart City Plan is being implemented. A Jaipur is nominated for World Heritage City status by JMC

Since the 18th century, Jaipur has housed 36 industries or *chattis karkha-nas*, which were designated as a City of Crafts and Folk Art as part of UNESCO's Creative City Network in December 2015. Indian Heritage Cities Network Foundation (IHCN-F) and DRONAH supported the local municipal body in achieving this designation. As part of this commitment, the JMC will be carrying out upgradation and urban conservation works in the streets of the walled cities that showcase a range of centuries-old crafts like *gota* work (golden thread) (see Figure 3.3), which would improve the socioeconomic status of the traditional craftspeople in the city (UNESCO 2015).

The conservation of the remaining nine bazaars of Jaipur was completed in 2019 through the Jaipur Smart City Plan. As per this scheme, the city had to prepare a vision plan engaging area-based plans, in consultation with citizens to define their own "smart" parameters.[6] Jaipur was one of the few cities where citizens collectively voted in favour of heritage planning within the walled city. The plan envisioned innovative and inclusive solutions that involved use of technology, information, and data to make infrastructure and services better, so as to enhance the quality of life of the residents. The areas within the walled city for the implementation of smart strategies included the main bazaar streets with the goal of preserving their historic

Figure 3.3 *Gota* work on cloth, a local craft of The Walled City of Jaipur. Source: Author

urban character through a series of interventions related to urban façade restoration and repairs, as well as façade illumination and general street improvements. Additionally, traffic control and provision of adequate physical infrastructure and services were also included as part of the implementation. The plan focused on carrying out urban conservation and adaptive reuse projects in the city incorporating all previous area specific management plans (such as the Jantar Mantar World Heritage Site Management Plan) and vision plans (such as the Jaipur Heritage Management Plan and Jaipur Master Plan 2025). The collaborative efforts of the local and national NGOs and municipal authorities have largely contributed to public awareness and local participation to preserve the "Walled City of Jaipur" for sustainable development. At the local level, the need to facilitate people's participation in management, urban governance, and technical capacity building, alongside raising awareness of the value of local heritage to ensure socioeconomic development, is necessary. Jaipur also faces urban pressures in terms of vehicular traffic and increasing real estate value that have taken a toll on the urban heritage and historic character of the city especially in recent years.

As per adopted Statement of Outstanding Universal Value (OUV) by the World Heritage Committee that is a guiding factor for its future management:

> The walled city has been recognized as a special area for heritage conservation under the Development Plan and shares the vision outlined

in the Jaipur Heritage Management Plan 2007. The Jaipur Heritage Management Plan has been implemented in various phases and synchronized with other plans from 2007–2017. Hence, in 2018, this Jaipur Heritage Management Plan was extended to include a comprehensive management strategy with an action plan for protecting the attributes and criteria contributing to the OUV. This Action Plan is intended to be used for the management and monitoring of the nominated World Heritage property. The action plan has been formulated taking into consideration the attributes identified under criterion ii, v, and vi; and how these attributes of the architectural style, gridiron plan, town planning principles, traditional house forms, iconic monuments, living traditions, and artistic works can be protected and safeguarded.

(UNESCO 2019)

The next steps are for the JMC to submit the Special Area Heritage Plan to the UNESCO World Heritage Committee by December 2020, along with providing a detailed inventory and monitoring indicators for all identified attributes.

Ajmer: heritage rejuvenation development and Augmentation Yojana

Ajmer is one of the 12 cities selected in the phase of HRIDAY (National Heritage City Development and Augmentation Yojana). The MoHUA assigned the role of City Anchors to NGOs and consultants in each city to work with the local government at each stage for envisioning and implementing the city's HRIDAY plan. DRONAH, along with ICLEI (International Council for Local Environmental Initiatives) – Local Governments for Sustainability and Indian Heritage Cities Network (IHCN) were selected as city anchors for Ajmer, to create the plan for the development of the city and coordinate between all levels of governments so as to propose a path for heritage-centric development.

Ajmer is the fifth largest city in Rajasthan with a population of around 551,360 in its urban agglomeration and 542,580 for the city, according to 2011 Census Data. The strategic location of Ajmer between the cultural regions of Marwar and Mewar[7] and its proximity to Delhi has made for a turbulent history resulting in complex layers of urban fabric. Home to the shrine of the Sufi Saint Khwaja Moinuddin Chishti, Ajmer is a leading centre for pilgrimage in India and attracts millions of visitors each year, not only Sufi Muslims but across faiths. Furthermore, Ajmer is also closely linked to the historic town of Pushkar, located just eleven kilometres away in the northwest. Pushkar is an ancient site of pilgrimage for Hindus, home to the only Brahma temple in the country. It is for these reasons that Ajmer is eminently suitable for re-development and upgradation under the HRIDAY.

Ajmer is also significant for its association with central events in Mughal–British history which forever changed the landscape of India. It was in Ajmer that Sir Thomas Roe, as ambassador of King James I of England, had his audience with Emperor Jahangir on 19th January 1616 CE, leading to the establishing of the British Raj in India through the charter of free trade granted to the East India Company by the emperor. It was also in Ajmer that Shah Jahan, on the death of Jahangir, proclaimed himself Emperor of India while returning from Udaipur and proceeding to Delhi in 1627 CE. In the nineteenth and twentieth centuries, the city emerged as a leading centre for higher education with premiere institutions such as the Mayo College, Sophia Senior Secondary School, and the Samrat Prithviraj Chauhan Government College established within the city. Furthermore, due to the location of Ajmer on key trade routes, it also became one of the first regions to become well connected via the railways which had a role to play in it becoming an industrial centre as well (Sarda, 1911) (Figures 3.4, 3.5).

While Ajmer has gone through some major infrastructure improvements in recent years (2010–14), specifically in the *Dargah* area under JNNURM and the Ana Sagar Lake area under the NLCP (National Lake Conservation Plan), its heritage areas still remain under major stress with increasing density, deteriorating historic fabric, and lack of visitor facilities. Pressure from growing tourism is also impacting the heritage areas as evident in an

Figure 3.4 View of Dargah Sharif, Ajmer. Source: Author

Figure 3.5 View of Baradari at Anasagar Lake. Source: Author

analysis of recent tourism trends in the city that showed an unusual growth of 47% in domestic tourists and 43% in foreign tourists from 2007–12, bucking the state-wide tourism trends, which have shown a decline for this period (City HRIDAY Plan Ajmer 2015, 2). As per the Annual Progress Report of Department of Tourism, Rajasthan, Ajmer recorded an annual increase of domestic tourists from 4.5 million to 4.7 million from 2015–18 (Department of Tourism, 2018–19). International tourists increased approximately from 36000 to 53000 in this same period. It should also be noted that while Ajmer itself shows an erratic increase in domestic tourism, there is a considerably slower growth rate of tourism in the nearby pilgrim town of Pushkar.

Existing governance mechanisms

The Ajmer Municipal Corporation (AMC) is in charge of providing services to the city, addressing grievances of the locals and executing development projects including infrastructure works within the municipal boundary which comprise the historic walled city. National schemes such as HRIDAY are also executed by the local municipal corporation supported by the HRIDAY city anchor. The ASI has 12 protected monuments and sites in Ajmer and Pushkar while the State Government of Rajasthan protects another 14 monuments and sites in the city. A number of national schemes including PRASAD, SMART City, and HRIDAY have already been implemented in the city. Another central scheme of the Ministry of Urban Development, AMRUT (Atal Mission Rejuvenation and Urban Transformation), to enhance infrastructure and green spaces in Indian cities, is currently being implemented in Ajmer.

Key aspects of the culture-based regeneration strategies

For cities like Ajmer, which fall under both HRIDAY and PRASAD, the government aims to ensure that the projects under both schemes are complementary and address concerns related to pilgrim tourism as well as sustained urban development of the historic city.

The HRIDAY scheme attempts to "preserve and revitalise soul of the heritage city to reflect the city's unique character by encouraging aesthetically appealing, accessible, informative and secured environment. To undertake strategic and planned development of heritage cities aiming at improvement in overall quality of life with specific focus on sanitation, security, tourism, heritage revitalization and livelihoods retaining the city's cultural identity" (Ministry of Urban Development 2015). The city's HRIDAY plan identifies and prioritises the main heritage areas in the city, analyses them in detail with respect to urban and tourism infrastructure, and proposes a conceptual plan for the areas in accordance with the overall vision of the city.

Ajmer's HRIDAY plan envisages the twin cities of Ajmer–Pushkar as a haven for pilgrims and visitors and a sustainable liveable environment for residents. The core heritage areas identified for rejuvenating include the Dargah precincts, Naya Bazaar, Ana Sagar Lake, Daulat Bagh or Subhash Udyan, Edward Memorial zone, Brahma temple, and Pushkar Lake. The City Infrastructure Development Plan intends to supplement the existing infrastructure gaps for these five culturally significant zones in terms of drainage, mobility, access, historic character legibility, and visitor facilities such as water fountains and toilets, to ensure that they transform into tourist-friendly areas reflecting iconic landmarks and cultural identity of the city.

The methodology adopted through the project first identified the key heritage assets in the city, followed by an analysis of the existing urban and tourism infrastructure within the identified heritage zones, and then improving and filling gaps in existing infrastructure, while conserving and restoring key heritage assets within their contextual urban setting. This would lead to an overall physical improvement as well as quality of life in the area. Livelihood improvement happens when there is creation of jobs in the service industry and tourism sector, enhanced sales, and need for service providers such as rickshaw pullers and tour guides. As a result of HRIDAY which prioritises heritage assets, there is a greater awareness of the heritage of the city among the locals as well. All the projects, however, do not fully engage the historic layers of the city, its varied culture and architecture, nor integrate well with users' needs and communities' associations with the spaces. The urban regeneration efforts ought to account for all these for sustainable urban conservation strategies.

The HRIDAY plan clearly outlines infrastructure gaps and heritage conservation initiatives in the form of a civil infrastructure development plan and further guides new development thus ensuring that the city retains its historic built character (and intangible cultural resources) into the future. The toolkit informs the local administration in monitoring any new development, additions, alterations, routine maintenance, and demolitions that may be undertaken in a historic context by private stakeholders, religious organizations, and such. Typically, planning decisions are taken after discussions with the HRIDAY National Empowered Committee (HNEC), MoHUA that comprise officials from all concerned departments as well as stakeholders, ensuring that there is no overlap or gaps in the proposed work and all schemes are implemented in tandem. For example, for the projects undertaken in the Naya Bazaar area, it has been ensured that all civil and electrical works are done in coordination with the electrical department and municipal corporation. Along with these works, concurrent façade restoration was also undertaken by the Rajasthan State Archaeology Department. This minimizes redundancy of efforts and disruption to traffic and everyday life in the project area.

While the current guidelines for HRIDAY are heavily focused on creating tourism and urban infrastructure for heritage assets in the city, one positive impact is the finalization of a "Heritage Toolkit" for designing signage and street furniture across the city. This toolkit is devised to ensure that the historic urban language of the city is maintained by a consistent design of streetscape furniture, lights, benches, dustbins, signage, and such that complement the city's historic context. Here it is critical to understand the cultural significance that shapes the essence of the city like the historical layers, associations, ritual processes, and activities that have – and continue to – shape historic sites. This inquiry into cultural significance was intended to be the primary focus of study of the walled city of Ajmer and nearby Pushkar, Rajasthan, a project supported by the Arts and Humanities Research Council (AHRC) and the Indian Council for Historical Research (ICHR) through a Research Networking Grant for "Cultural Heritage and Rapid Urbanisation in India". The research was jointly conducted by Cardiff University, SPA Bhopal, and DRONAH Foundation in 2016 (DRONAH 2016a).

Through the course of the research, prototypes of digital tools containing images and information about the city, its history, and heritage were created to allow reflection, exchange, and dialogue among different stakeholders. In a direct way, the research addresses the tangible and intangible cultural heritage in the context of rapidly urbanizing India. It integrates built heritage and prioritizes heritage conservation within urban planning. In order to create an overview of the city as a whole, a city-level digital interactive map of the historic areas of Ajmer was prepared and is available on the DRONAH Foundation website. As part of the digital mapping, historic structures and routes were marked in the context of neighbourhoods showing urban grain and important nodes. Local residents were interviewed as a participatory approach to the project proposals (DRONAH 2016a). Recommendations for adaptive reuse of private historic structures emerging from the research project are intended to support ongoing HRIDAY works in the area. For example, specific recommendations were made for pedestrianization and reuse of the Gol Piau Chowk as a public space along the heritage walk. The owner of Gota Factory Haveli was shown how his *haveli* could be opened up for visitors interested in historic paintings and weaving of *gota* (golden thread) without impacting his work.

Lessons and conclusion

Implementation of urban heritage-related projects in Indian cities is challenging as they are often located in older and congested areas facing developmental pressures, increasing traffic, escalating real estate values, and multiple ownership of private properties. Additionally, multiple authorities are usually involved during project implementation, which often delays the process due to lack of coordination among them. Community involvement and creating awareness at the time of project execution is important to ensure minimal resistance and maximum support for the proposed interventions. Successful implementation strategies and ongoing maintenance are key factors to the success and long lasting impact of urban heritage-related projects in a city. Insights from public engagement as part of the participatory design approach or through NGOs should inform ongoing HRIDAY projects.

Efforts have been made in the past few years to incorporate heritage management initiatives at the city level with the policy interventions by the Government of India. However, such city-level visions finally need to be realized in tangible and sensitive projects that are implemented to improve the living environment for local citizens in historic city cores. These projects need to be sustainable, with a focus on a community-based approach. It has been realized that for a successful and effective implementation of any plan in a historic area, support of various interest groups is beneficial, be it the government authorities, the residents, NGOs, or visitors. Besides policies, data collection and documentation of heritage, conservation, and urban renewal, interpretation and heritage awareness as well as heritage valuation play a critical role for longevity and sustenance. The relevance of a contextual framework of ownership, significance, and economic potential of heritage should be understood, and planning should reinterpret and adapt the historic character to contemporary realities.

The sustainability of urban conservation projects also resides in anchoring them fundamentally to a city's vision plan along with an inclusive participatory approach while implementing conservation work. The HRIDAY programme is in the process of reviewing the policies and guidelines for revisions based on the ongoing works in 12 cities. The first document for recommendations was prepared by MoHUA in association with ICOMOS (International Council on Monuments and Sites) and CEPT, Ahmedabad. It was released in December 2017 at the ICOMOS General Assembly in New Delhi. It emphasized the gaps in socioeconomic aspects of selected projects along with capacity building of local municipal engineers in undertaking conservation works. Subsequently, MoHUA is planning more such review documents, considering that HRIDAY projects in all 12 cities will be completed by the end of 2020.

Considering the repository of historic cities in India, the conservation and management of urban heritage is an immense task for the future. The challenge is much larger compared to what has been achieved in western cities because of the complex layers of history, living heritage, craftsmanship combined with other urban sectoral issues of services, mobility, and infrastructure. The government initiatives in the last decade have recognised the significance of historic Indian cities. The inscription of Ahmedabad and Jaipur as World Heritage Cities in 2017 and 2019 respectively, has further supported the international acknowledgement of the vast urban heritage in India. Inclusion of visionary heritage plans in government schemes is a welcome enterprise to boost past efforts of the conservation community, practitioners, academicians, and NGOs in India who were proactively pursuing this trajectory. Besides the 12 HRIDAY cities, a number of Smart City Plans have also incorporated heritage conservation of historic cores. Jaipur and Ajmer, besides other Rajasthan cities like Udaipur, are precursors in this movement and learnings from these can provide useful recommendations for future interventions in other historic cities in India.

Notes

1 The State Government of Maharashtra enforced the Heritage Conservation Act in 1995 by amending the D.C. Regulations. In this regard 615 structures/ precincts were listed as heritage structures/precincts and graded into 3 categories as per their significance. (Mumbai City Development Plan 2005-2025) along with formation of a Heritage Committee. A similar process was followed in Hyderabad. Based on these examples, Ministry of Environment and Forest, India created Model Heritage Bylaws to be used by city municipalities. These were further modified and improved by the Ministry of Urban Development (Ministry of Housing and Urban Affairs since 2015) in 2014.

2 HRIDAY scheme was launched with a focus on holistic development of heritage cities and by November 2018 (4 years of the project), the scheme had been implemented in 12 cities in India including Ajmer. The scheme focuses on development of heritage infrastructure like water supply, sanitation, drainage, waste management, approach roads, footpaths, street lights, tourist conveniences,

electricity wiring, landscaping etc. around heritage assets identified/ approved by the MoHUA, Government of India and state governments.

3 World Monuments Fund (WMF), Asian Development Bank (ADB), United Nations Educational, Scientific and Cultural Organization (UNESCO), Archaeological Survey of India (ASI), Ministry of Housing and Urban Affairs (MoHUA), State Government of Rajasthan, royal patrons, local municipal agencies, and local residents.

4 The Department of Local Self Government is the controlling department of all the municipalities for administrative purposes and monitors and coordinates functioning at the state level for all the municipal bodies of the state of Rajasthan.

5 Chowkri Modikhana is one of the historic chowkri in Jaipur, which gets its name from the historic association with the Modis or the trader community. The areas has *havelis* (elaborate residential units), temples, public buildings, museums and craft areas which are a unique representation of the living heritage of Jaipur.

6 As per MoHUA vision, the concept of Smart City varies by city and country, depending on the level of development, willingness to change and reform, resources and aspirations of the residents.

7 Marwar, present day Barmer, Jhalore, Jodhpur Nagaur, Pali and parts of Sikar districts of the present state of Rajasthan and Mewar, present day Bhilwara, Chittorgarh, Rajsamand, Udaipur districts are two the cultural regions within Rajputana.

References

Archaeological Survey of India, Jaipur Circle. http://www.asijaipurcircle.nic.in

Atal Mission Rejuvenation and Urban Transformation, Ministry of Urban Development, Government of India. Accessed October 2, 2017. http://amrut.gov.in/

Bhattacharya, M. 2008. *The Royal Rajputs: Strange Tales and Stranger Truths*. New Delh: Rupa and Company.

Department of Tourism, Government of Rajasthan. 2018–19. Annual Progress Report. Government of Rajasthan. http://www.tourism.rajasthan.gov.in/content/dam/rajasthan-tourism/english/others/tourism-department-annual-progress-report-2018-19.pdf

DRONAH. 2016a. AHRC-ICHR Research Networking Awards on Cultural Heritage and Rapid Urbanisation in India on the *Theme* 'Public Spaces and Urban Planning'. Accessed May 25, 2020. https://www.dronahfoundation.org/ahrc-ichr-research-networking-awards-on-cultural-heritage-and-rapid-urbanisation-in-india-on-the-theme-public-spaces-and-urban-planning/

DRONAH and JVF. 2009. *Façade Control Guidelines, Chowkri Modikhana, Jaipur Walled City*. Jaipur: Jaipur Municipal Corporation. Unpublished report.

Gole, Susan. 1989. *Maps and Plans of India*. Delhi: Manohar Publishers.

Hardy, Adam. 2007. *The Temple Architecture of India*. Wiley, Chichester.

Hooja, Rima. 2006. *A History of Rajasthan*. New Delhi: Rupa & Co.

Hooja, R., Hooja, R. & Hooja, R. 2010. *Constructing Rajpootana – Rajasthan*. Jaipur: Rawat Publications.

Imperial Gazetteer of India: Provincial Series Rajputana. 1989. Pune, Maharashtra: Usha Publications.

Jain, K. C. 1972. *Ancient Cities and Towns of Rajasthan: A Study of Culture and Civilization*. Delhi: Motilal Banarsidass.

Jain, Shikha (ed.). 2005. *Princely Terrain: Amber, Jaipur and Shekhawati*. Gurgaon: Shubhi Publications.

Jain, Shikha. 2010a "Jaipur as a Recurring Renaissance", in *New Architecture and Urbanism*, ed. Deependra Prasshad. Newcastle upon Tyne, UK: INTBAU, Cambridge Scholars Publishing, 60–68.

Jain, Shikha. 2010b. Walking into the Microcosm of Jaipur: A Concept Paper. Accessed June 30, 2018. http://unesdoc.unesco.org/images/0019/001921/19211 3e.pdf

Nath, Aman. 1996. *Jaipur, The Last Destination*. London, UK: I.B. Tauris & Co, Ltd.

UNESCO. 2019. World Heritage Listing for Jaipur City, Rajasthan. Accessed May 25, 2020. https://whc.unesco.org/en/list/1605/

4 Tools for heritage advocacy in Lucknow

Active civic engagement and public interest litigation

Ashima Krishna

Introduction

Activism and advocacy are not new in India. Both social and political activism[1] and advocacy[2] have been part of the country's fabric for at least two centuries, bringing about major changes (Shaikh 2012). In the postcolonial decades, and especially since 1947, urban and environmental activists and advocates have brought about policy and legislative changes (for more on this see Chainani 2007a, 2007b, 1982; Mehrotra and Lambah 2004). They rely on public interest litigation (PIL) that has had a huge impact on slowing down, if not stopping, actions that are deemed detrimental to public welfare (for examples see Follmann 2015; Appadurai 2002; Pai 2004). Environmental protection, in particular, has been sought through what has been called "judicial activism", using PIL as an effective tool, although at times detrimental mega urban projects have unfortunately forged ahead due to political support (Follmann 2015, 221).

Since the 1970s PIL has been an integral part of the Indian appellate judiciary[3], distinct from its American and Canadian counterparts (Bhuwania 2014, 314). Despite its rather fraught beginnings, PIL has increasingly become a tool to affect policy change and social reform. PIL cases are typically filed by appellate lawyers in the High Court or the Supreme Court, thus allowing judges to provide "pious homilies" and guidelines in their judgements (Bhuwania 2014, 322). It has been found that PIL is also used as a populist tool by the courts to render judgement on matters of public importance (Bhuwania 2014, 322). One of the most famous cases of the use of PIL as an activism and advocacy tool was by environmental lawyer M.C. Mehta in the 1990s in his fight to save the River Ganga and the Taj Mahal from excessive industrial pollution (Kalshian 1996). Since then, Mr. Mehta has continued to file PIL cases in order to compel actions by state and central governments that can preserve and protect the Taj Mahal and its environs (for more on PIL cases see Supreme Court of India 1996b, 1996a, 1997, 1998, 2000, 2003b, 2003a, 2006, 2007, 2010).

More recently, youth in India have also used social media to bring about social change (Rishi, Bhanawat, and Bajaj 2012). Contemporary activism and advocacy efforts, therefore are expanding beyond traditional roles to encompass several ways in which the public is able to engage with and advocate for issues they are passionate about. This chapter examines the work of one heritage advocate in the city of Lucknow, Syed Mohammad Haider Rizvi (henceforth Mohammad Haider), and focuses on the efforts to manage and preserve historic sites located in the heart of the city through active civic engagement and PIL cases. These tools take on a particularly important role in cities like Lucknow that have rich architectural and cultural heritage, but hardly any local policy framework to protect that heritage. Even central and state level legislations are often either outdated or ineffective, as several cases discussed in this chapter demonstrate.

Heritage advocacy tools in Lucknow

Lucknow has had a long history of advocacy and civic engagement. *Jalsah-e-Tahzib*[4], a voluntary cross-community alliance established in February 1868, comprised a small group of middle-class educated Hindus and Muslims whose mission was to promote the city's culture in the post-1857 era. The group, however, while inclusive of religion and caste, tended to exclude based on education and refinement; consequently, they comprised the administrative elites of Lucknow society: "government servants, members of the legal profession, and men belonging to the vibrant world of vernacular print and journalism" (Stark 2011, 5). Their success was predicated on "negotiation and collaboration with the state" (Stark 2011, 29). This group aimed to shape the intellectual and cultural life of Lucknow through advocacy and outreach – in contrast to *Anjuman-e-Hind* (also known as the British Indian Association or BIA, created 1861), which comprised the landed gentry called *taluqdars* who were encouraged by the British to lead and participate in the public affairs of Lucknow, particularly in the wake of the First War of Independence in 1857 (henceforth The War of 1857) (Stark 2011). The BIA has been involved with charitable, cultural, social, and educational efforts across the city for nearly two hundred years, and has been the custodian of several heritage properties in the city (for more on the BIA see Krishna 2014).

More recently, heritage and urban advocates in Lucknow have had some success in championing *for* the protection of urban heritage, and *against* projects that could have adverse impacts on the city's built legacy. In 2010, for example, many well-connected residents of the city banded together to form *Connect Lucknow*, an advocacy group that helped facilitate the revitalization and restoration of the historic Hazratganj market street. Connect Lucknow has some parallels with *Jalsah-e-Tahzib*; their members are largely well-educated, have political connections, and include members of major newspapers in the city. These connections helped the group leverage significant public support and political pressure to ensure the completion of the revitalization project. Additionally, given the strong economic interests of the businessmen from the area, city and state agencies had to ensure that the project was timely and successful to not incur financial loss for businesses located along the market street (for more on this see Krishna 2014, 2016, forthcoming).

Other areas in the city (both designated and not designated), unfortunately, have not enjoyed the same level of attention, and have been largely dependent on heritage advocates to fight for their cause. Given the high volume of Nawabi-era built heritage in the city, the majority of such advocacy efforts have been focused on the city's sizeable Islamic heritage, including several iconic *imambara*[5] complexes.

Active civic engagement

Mohammad Haider, a lawyer, began advocating for heritage causes in 2008, a few years after he was appointed a joint *mutwalli* (caretaker) by the Shia Waqf Board to manage the Sibtainabad Imambara, which is also known as the Amjad Ali Shah Mausoleum (see Figure 4.1). Together with Syed Imtiyaz Alam, Haider has been responsible for the management and maintenance of the mausoleum since 2000. Both *mutwallis* inherited a long history of preservation challenges at the *imambara*. As far back as the aftermath of the War of 1857, the *imambara* repeatedly faced conflicts of ownership between the then British government, descendants of the last *nawab*, Wajid Ali Shah, and city agencies (for more on this see Krishna 2014, forthcoming). In the 1860s it was temporarily used as a church for about two years while Christ Church was being constructed (Hay 1939, 167).

In December 1918, the *imambara* was acquired by the Lucknow Improvement Trust (today the Lucknow Development Authority), and declared a public property under the Land Acquisition Act, and soon after, in November 1919, it was declared a "protected monument" under the Ancient

Figure 4.1 Over decades, there has been significant unchecked construction activity in and around the residential quarters along the periphery of the mausoleum complex, seen in the extreme left and right of the image. Image courtesy Michael Tomlan, 2013

Monuments Preservation Act of 1904. The two gateways to the property and surrounding buildings were added to the listing in 1924 (Llewellyn-Jones and Kidwai 2011; Haider 2012). In 1921, another part of the complex was acquired by the Lucknow Improvement Trust. Subsequently, they let out the adjoining spaces in the complex (as shown in Figure 4.1) for residential and commercial use (Haider 2012). By 1922, sixty-six quarters within the inner court of the mausoleum complex were allotted to Anglo-Indians by the local *Nazul* office (Llewellyn-Jones and Kidwai 2011). Several of the quarters are still occupied by the descendants of the original occupants, and are referred to in the next section.

In the period after Indian independence, the Directorate of Agriculture and the Census Office of the Government of Uttar Pradesh's offices were housed in the *imambara* building for several decades. Once they vacated the building, it was rented to a furniture company, which ran its business from within for over thirty-five years. Finally, at the turn of the century, the *imambara* was declared a *waqf*[6] property and handed over to the Shia Waqf Board for maintenance and management (Llewellyn-Jones and Kidwai 2011; for more on the Shia Waqf Board, see Krishna 2014). By this time, the site had undergone decades of neglect and abuse both in the *imambara* building, and in the adjoining quarters and gateways as shown in Figure 4.2.

Figure 4.2 Significant construction in and around the inner gateway to the mausoleum complex has rendered it almost unrecognizable (the mausoleum is visible through the gateway in the background). Image courtesy Michael Tomlan, 2013

Engaging the public agencies

Despite this property being a nationally protected monument, little was done over the years by the ASI and state and local agencies. Over decades, the Lucknow Circle of ASI repeatedly wrote letters to the local administration complaining about the irreparable damage being done to the historic structure by various legal and illegal occupants. In several instances, the ASI also registered complaints with the local police[7] in an effort to stop illegal construction within the complex. Unfortunately, despite several complaints, there was little to no action taken by the local administration for several years (Mohammad Haider, lawyer and heritage advocate, in discussion with the author, March 2012). Consequently, over the years, Alam and Haider have often had confrontations with the ASI, Lucknow Development Authority (LDA), Lucknow Municipal Corporation (LMC), and the area residents. Haider, in particular, has been actively engaged with various civic agencies and has attempted to engage with the local residents.

There is a long history of neglect by the ASI and apathy and contestation between the ASI and state and city agencies (Ahmed 1998). Haider recognized these challenges, and began to advocate for better conditions within and around the *imambara*. Beginning in 2008, he wrote several letters and emails to the Director General of ASI in New Delhi, noting several problems. These included "blocked access to outer gate of the complex by construction made by a clothing store; encroachments in and around inner and outer gates; encroachments made by residents of quarters along the boundary; and use of the complex as a car park" (Krishna 2014, 260). He also noted that the ASI's intervention was urgently needed because the Lucknow Development Authority (formerly Lucknow Improvement Trust) had plans to transfer the deeds to the occupants of the adjoining residential quarters for a nominal fee – this would endanger the already at-risk historic integrity of the complex. It was felt that this step would also impede any future work by the Shia Waqf Board and the ASI in restoring the *imambara* (Haider, in discussion with the author, March 2012 and January 2016). In September 2000, the Lucknow Circle of ASI requested the local administration to help in dealing with the occupants (both legal and illegal) of the residential properties, as well as unauthorized construction within the *imambara* complex, as it was impeding their ability to carry out conservation work. Little to no action was taken in the following years.

By August 2008, the ASI began repairs on the collapsed roof of the *imambara*, after its condition was brought to their notice by the *mutwallis*. The ASI again asked the local administration to remove the illegal encroachments in the area, and again, no action was taken. In May 2009, the ASI filed a complaint in the Hazratganj Police Station against more unauthorized construction work being carried out within the premises. The two *mutwallis* of the complex agreed with the complaint, eventually leading the ASI to issue a show cause notice to the offenders; again, however, without sustained follow-up (Haider, in discussion with the author, March 2012). Repeated complaints were made by both Alam and Haider, but to no avail. The ASI also attempted to acquire the remaining complex from the LDA, and encouraged the agency to resettle the occupants of the residential quarters elsewhere in the city. The acquisition, however, was not successful. Consequently, the ASI continued with their repair work on the damaged portions of the *imambara*; they were, however, unable to continue with the extended repairs of the complex without the cooperation of the local administration. That same year, the ASI was unsuccessful in stopping a local commercial establishment from undertaking unauthorized construction along the outer gateway to the complex as shown in Figure 4.3 (Haider, in discussion with the author, March 2012). On April 2 2020, the outer gateway eventually collapsed under the weight of excessive unauthorized construction.

There have been inter-agency challenges as well. As mentioned earlier, in 2010, Hazratganj, an upscale market street in the heart of Lucknow and adjoining the *imambara* complex, was revitalized through a public-private partnership between city and state agencies and local businessmen (Krishna 2014, 2016, forthcoming). During the project, unauthorized construction work was again carried out in very close proximity to the complex and its gateways[8], not authorized by the ASI as per federal legislation, especially

Figure 4.3 Extensive encroachment and construction activity has taken place, relatively unchecked, along the outer gateway to the mausoleum complex. Source: Author, 2013

Sections 20A and 20B of the Ancient Monuments and Archaeological Sites and Remains (Amendment of 2010). According to Sections 30A and 30B, the offence was punishable by imprisonment for up to two years and monetary fines.[9] Despite the ASI's objections, the work, however, was successfully completed by local and state agencies.

Over the years, the *mutwallis*, and Mohammad Haider, in particular, have been able to successfully get the ASI to act in the best interests of the protected monument through repeated advocacy campaigns, writing letters, and seeking assistance from print media. Their efforts finally bore fruit, and after several decades of neglect, all unauthorized uses were removed from the *imambara*'s vicinity in 2010. The ASI also undertook major conservation work to restore the interior and exteriors of the *imambara* in 2011. This work was overseen by the *mutwallis*, eventually allowing the Shia community to finally hold religious events like *majlis* in the *imambara* (Haider, in discussion with the author, March 2012).

The ASI's efforts at the site have been relatively sporadic. Haider has, on several occasions, requested them to create a comprehensive management plan to help preserve the complex; however, the ASI continues its piecemeal efforts at the site. Haider has also offered to help in fundraising from local political funds; however, the work has been piecemeal (Haider, in discussion with the author, January 2016).

Haider also sought political support in his effort to save the site. In mid-2015 he petitioned the then Uttar Pradesh State Minister for Urban Development to intervene and get repair work carried out at the *imambara* complex. Haider's process of engaging with civic agencies has included writing letters, following up with relevant agencies about those letters, and filing a request through the Right to Information Act (RTI)[10] with those agencies in connection with the letters written. With every letter, Haider prevailed upon the agencies to facilitate corrective action at the earliest. The follow-up and RTI requests helped Haider monitor the progress on the resolution of issues, and also allowed the work to be more transparent.

Over the years, Haider has also campaigned for local, state, and central agencies to designate and protect at-risk sites, particularly those that meet the ASI's one-hundred-year rule (Haider, in discussion with the author, January 2016). Haider filed another RTI application with ASI's Lucknow Circle and found that they had not designated or protected any new sites since 1920, primarily because the ASI found the process to be cumbersome and onerous (TNN 2013).

One important tool employed by Haider in dealing with different kinds of civic agencies, special interest groups, and community members has been through extensive documentation of all the appeals and requests made to various government agencies to establish a legal basis for follow-ups. Typically, he also undertakes extensive research before approaching an advocacy project, including reviewing all public documents and orders, and reading all available material, like books, research papers, the gazetteers, and other primary sources related to a site and its issues. This work is done to properly prepare for every court appearance related to PIL and other heritage-related court cases, so that he has answers to any questions asked by judges (Haider, in discussion with the author, January 2016).

Engaging the community

From the beginning, Haider has attempted to employ a collaborative approach in working with the residents living within the Sibtainabad Imambara complex. He sought cooperation from the Sibtainabad residents by writing an open letter in 2008, asking all residents to meet him at a particular date and time. Almost ninety percent of the residents came for the meeting, discussed all of their various concerns, and worked with Haider to come up with amicable solutions. Before the meeting, most residents suspected that Haider and Alam both wanted to have them evicted. With the in-person meeting, however, Haider assured the residents that his goal was to have the designated structure be appropriately restored by ASI, clarifying that his PIL made no mention of the residents of the surrounding residences (Haider, in discussion with the author, January 2016).

Haider and the residents also agreed that the garden fronting the mausoleum should be maintained, and that the local children should refrain from playing sports in the area, which had been damaging the flora. Haider sought their cooperation to engage all residents and earn their goodwill. He felt that if the residents took pride in their heritage, they would be more likely to safeguard it.

The area residents were involved with Haider in identifying the problems of the area, develop corrective actions, crafting appeals to LMC and the local police, coordinating with the LDA for better ingress and egress to the site, repair of interior roads, and removal of motor garages from the complex. This collaborative approach helped Haider and Alam to build better relationships with the residents of the complex. In turn, the residents (who are predominantly non-Muslim), were more open to the Islamic religious events held within the *imambara* (Haider, in discussion with the author, January 2016). While the collaborative approach has had relative success at the Sibtainabad Imambara complex, it hasn't worked elsewhere, primarily because of vested interests among some of the stakeholders involved: encroachers of heritage areas in the city, contractors who carry out repairs, henchmen of nefarious government officials, as well as local politicians. These kinds of stakeholders do not want to have a dialogue with heritage advocates like Haider and others. In those cases, Haider has resorted to PIL and court cases.

Public Interest Litigation (PIL)

Haider's heritage advocacy has involved other sites in the city as well, primarily through PIL. In 2013, Haider filed a PIL with the Lucknow Bench of the Allahabad High Court, noting the dismal state of several protected sites in the city, and asked the court to ensure that the relevant central and state agencies maintain and preserve them. The PIL requested the court create a committee of experts who could help identify the problems faced by the ASI and the UP State Archaeology department in preserving and maintaining historic sites, and to suggest measures to curb encroachment and defacement of these sites (Chief Justice's Court 2013). On October 10, 2013, the High Court issued an order to constitute a committee chaired by the Commissioner of Lucknow, with the Registrar of the Court acting as its Convenor. The court indicated that the committee should consist of: the District Magistrate of Lucknow, the Senior Superintendent of Police, an Additional Municipal Commissioner nominated by the Municipal Commissioner, Secretary (Lucknow Development Authority), Jaideep Narain Mathur (a senior advocate), Syed Mohammad Haider Rizvi (the petitioner), Vipul Varshney (convenor of INTACH Lucknow), and the Superintending Archaeologist of ASI's Lucknow Circle (Lucknow Bench of Allahabad High Court – Court No.2 2013c). Following the October 10 order, on October 23, 2013, the High Court recommended that the State government should form "Model Regulations for conservation/preservation of heritage sites, (both natural and man-made)" (Lucknow Bench of Allahabad High Court – Court No.2 2013d).

In an order dated November 27, 2013, the High Court directed the ASI and the District Magistrate of Lucknow to "take appropriate steps for removal of unauthorized/illegal constructions/encroachments in and around the four protected monuments situated in the Qaiserbagh area, Bada Imambara, and Roomi Gate". In this order the court also asked the committee to deliberate on the model regulations drafted by the state government under Section 56(2) of the U.P. Urban Planning and Development Act, 1973 (Lucknow Bench of Allahabad High Court – Court No.2 2013b).

Haider also discovered that Husainabad Allied Trust (HAT), the principal stakeholder of Husainabad Imambara (Chhota Imambara), was carrying out inappropriate restoration work on the site's façade (Figure 4.4). He wrote numerous letters to HAT and ASI in an effort to halt the egregious actions. In 2013, after several unsuccessful attempts at contacting the agencies involved, Haider filed a continuing motion to add to his existing PIL with the Lucknow Bench of the Allahabad High Court, requesting halt of use of inappropriate cement in repair work being carried out at the Chhota Imambara. By the middle of 2013, parts of the outer gate of the *imambara* had also fallen (TNN 2013). In March 2014, the High Court directed the ASI to carry out repair work on the fallen gate. The ASI filed a petition stating that the gate was out of their purview, and showed the protection document from 1920 as proof. Haider, the petitioner, pointed out that according to the definitions provided in Section 2(a) clauses i, ii, and iii, of the Ancient Monuments and Archaeological Sites and Remains (AMASR) Act of 1958,

Figure 4.4 Husainabad Imambara, or Chhota Imambara as it is popularly known, has undergone insensitive and ignorant restoration in the last few years by its principal stakeholder, Husainabad Allied Trust. Image courtesy Michael Tomlan, 2013

the definition of "ancient monument" also included adjoining portions of land, and any fencing or covering, which established that the gate was part of the site, and therefore under the purview of the ASI. The court asked the Senior Superintendent Archaeologist of ASI's Lucknow Circle, the District Magistrate of Lucknow, and the Principal Secretary/Secretary of the state's Department of Culture to ensure the protection of the gate (Lucknow Bench of Allahabad High Court 2014).

In 2014, the residents of the Sibtainabad Imambara area hired a renowned lawyer and filed an application against Haider's 2013 PIL. They argued that Haider's actions were personally motivated, and not in the interest of the public. Their concern again stemmed from the misconception that Haider wanted to have all the residents of the quarters evicted. Haider again engaged with the community and reiterated that he was concerned about the monument, and the inner and outer gates, and had no intention of having the residents evicted (Haider, in discussion with the author, January 2016). Eventually, at the hearing, the court dismissed the residents' application, because the judges argued that Haider's PIL clearly stated that it was concerned with protecting designated historic sites around the city, and did not mention the eviction of the residents of the Sibtainabad Imambara complex (Lucknow Bench of Allahabad High Court – Court No.2 2013a).

In September 2015, the Court adjudicated the matter of inappropriate conservation work, and asked HAT to stop construction, and instructed the ASI to step in and take corrective measures (Frederick 2017; Husain 2017). This resulted in an order from the High Court that led to the creation of an expert committee to

> examine the extent of damages caused by the recent restoration work by the HAT and to suggest further remedial measures [and] to evaluate the matter of protection of outer gates of the Chhota Imambara Complex in compliance of the order of the Hon'ble [Allahabad] High Court, Lucknow Bench dated 22.05.2014.

The committee found that the work carried out by HAT had done irreparable damage to the historic fabric of the structure, but removing it would cause further damage (Lucknow Bench of Allahabad High Court 2015). By July 2016, Haider again approached the HAT, ASI, and the district administration because very little work had been done, and it was in contempt of the previous order of the court (TNN 2016). By August 2018, there had been no further progress, largely due to the district administration's inability to release funds for the work (TNN 2018).

Haider also filed an RTI application with ASI to find out about funds spent between 2000 and 2012 on nationally designated sites. Subsequently it was found that just over half of all nationally protected sites in the city received funds for preservation work in the twelve-year period (TNN 2013). Haider also filed another PIL that mentioned the Asafi Imambara (Bada Imambara) and the encroachment around it. Haider requested in his PIL that all the encroachers be given an alternate residence elsewhere so that they have security of tenure, otherwise they may return to the site. As a consequence of this PIL, all the encroachers were relocated to a public housing site near Alamnagar under the then state government's Lohia Awas Yojana (Haider, in discussion with the author, January 2016).

In 2018, Haider's advocacy efforts bore fruit, and the National Monuments Authority selected Sibtainabad Imambara as one of twelve sites from the Central region to receive site-specific heritage byelaws. In November 2018, Sibtainabad Imambara became the first site in the country to get its own publicly available bylaws, which included detailed design guidelines on the range of the types of construction activity allowed in the area, the height, and the materials as well (Lalchandani 2018). This is an incredible development for the historic precinct, and illustrates the effectiveness of continued advocacy. While Lucknow still does not have heritage byelaws for the entire city, site-specific byelaws for a much-contested site like the *imambara* can help spur on a more city-wide effort at regulating historic sites. The High Court's directives to city and state agencies, while still in progress, are also a step in that direction, and were a result of the PIL filed by Haider.

Lessons for preservation practitioners

The decade-long struggle by heritage advocates in the city of Lucknow shows that there are several tools available to the community to assist with the protection of their heritage, especially in the absence of local policies and ordinances. Meticulous documentation of all correspondence with various agencies is very important to maintain records for future reference, particularly if advocates plan on using PIL as a heritage advocacy tool. Haider's success shows that heritage advocates also need to conduct thorough research and collect appropriate data before approaching the media, the public, or the courts. Anyone who wants to do heritage advocacy needs to do their research and understand the site, its significance, and all documentation pertaining to it, including any local, state, or central orders, directives, and designations.

It is also important for heritage advocates across the country to be familiar with the legal provisions at central, state, and local levels. In the event that there is a perception of any detrimental actions that would have an adverse impact on a historic or cultural site, then heritage advocates should be able to seek legal recourse, especially as most heritage sites are often vulnerable. Heritage advocates also need to work with local residents and seek collaboration and cooperation from community members. Getting to know the community being advocated for, or being affected by the advocacy is important, especially since everyone's motivations may not be heritage-centric.

Social media and print and television media are also incredibly important tools for a heritage advocate. Social media can help disseminate information faster than conventional means, and also mobilize support. Print and television media have been tools for advocacy for decades, and continue to do so. Most newspapers have a culture department, and they have journalists dedicated to finding stories in that field. Haider has, over the years, cultivated relationships with journalists from different newspapers, allowing him access to a very important tool. This has helped give a voice to all the various causes close to Haider, and has put his efforts in the spotlight, garnering further public support.

Indian cities have a wealth of architectural and cultural heritage that is often at risk, especially if it is not designated as a monument at the central or state levels. Even if sites are designated, a paucity of resources and viable enforcement mechanisms put them at risk to unchecked construction activity, encroachment, and development. In the absence of strong mechanisms at local, state, or federal levels, heritage activists and advocates are surrogates for the agencies involved with cultural heritage and its conservation. In many cases, like Shyam Chainani's efforts in Maharashtra, continued activism and advocacy led to strong local and regional legislation to protect heritage (Chainani 2007a, 2007b). Second-tier cities like Lucknow are still struggling to protect their heritage in the face of strong development pressures, government apathy, and lack of funding. Continued efforts by persistent advocates like Mohammad Haider, as shown in this chapter, however, have brought about targeted changes, and have made a significant positive impact on the cultural heritage of Lucknow. In the absence of legislative frameworks, heritage activists and advocates elsewhere could employ similar tools to champion for the protection of architectural and cultural heritage.

Notes

1 Activism is defined as the act of championing a cause (Carr and Sturdy Colls 2016, 703), often from outside the prevalent system.
2 Advocacy is also defined as an act of championing a cause, however, it is often seen as doing so from within the system, using the tools available.
3 The Indian judiciary comprises three levels: the first is the Supreme Court, based in Delhi; the second level comprises the high courts in every state, and they function as appellate judicial systems; the third level comprises the district courts. In India, the three levels of judiciary operate as "a single hierarchy and administer both state and federal laws" (Bhuwania 2014, 317).
4 *Jalsah* means "gathering" or "assembly" in Urdu; *tahzib* can be interpreted as "refinement", "culture", or "etiquette" (Stark 2011, 5).
5 An *imambara* is an Islamic mausoleum.
6 A *waqf* is an Islamic charitable endowment made in dedication to God by a donor (*waqif/wakif*) to allow for the donation of movable and immovable property in the name of God, especially to allow Muslims to benefit from its revenue (Sajid Shamim, Additional General Manager, Waqf Development Corporation, in discussion with the author, September 2011).
7 ASI needs the assistance of local administration and local police departments in enforcing the federal law, as well as acting to stop encroachments and actions that can cause adverse impacts on designated sites.
8 The legislation prohibits any unauthorized construction within 100m of a protected monument.
9 The Lucknow Circle of ASI also took out a public notice in newspapers regarding the 2010 Amendment to the AMASR Act of 1958, including the condition that no projects, including public projects could be carried out within the "prohibited areas".
10 RTI, or Right to Information Act was adopted by the Indian Parliament in 2005 and went into effect in October of that year. Similar to the Freedom of Information Act in the United States, the RTI Act gives citizens the right and access to information from public agencies. Unlike its American counterpart that applies only to the federal government, however, RTI applies to state and local governments as well (Roberts and Roberts 2010).

References

Ahmed, Farzand. 1998. "As Legendary Monuments in Lucknow Decay and Collapse, Callous Officials Start Blame Game." *India Today*, September 1998. https://www.indiatoday.in/magazine/heritage/story/19980928-as-legendary-monuments-in-lucknow-decay-and-collapse-callous-officials-start-blame-game-827113-1998-09-28

Appadurai, Arjun. 2002. "Deep Democracy: Urban Governmentality and the Horizon of Politics." *Public Culture* 14 (1): 21–47. https://doi.org/10.1215/08992363-14-1-21.

Bhuwania, Anuj. 2014. "Courting the People : The Rise of Public Interest Litigation in Post-Emergency India." *Comparative Studies of South Asia, Africa and the Middle East* 34 (2): 314–335. https://doi.org/10.1215/1089201X-2773875

Carr, Gilly, and Caroline Sturdy Colls. 2016. "Taboo and Sensitive Heritage: Labour Camps, Burials and the Role of Activism in the Channel Islands." *International Journal of Heritage Studies* 22 (9): 702–715. https://doi.org/10.1080/135272 58.2016.1191524

Chainani, Shyam. 1982. "The Role of Environmental Groups : A Bombay Experience." *India International Centre Quarterly* 9 (3): 268–281.

Chainani, Shyam. 2007a. *Heritage & Environment: An Indian Diary.* Mumbai: Urban Design Research Institute.

Chainani, Shyam. 2007b. *Heritage Conservation, Legislative and Organisational Policies for India.* New Delhi: INTACH.

Chief Justice's Court. 2013. "MISC Bench No. 3173 of 2013; Order Dated April 17, 2013."

Follmann, Alexander. 2015. "Urban Mega-Projects for a 'World-Class' Riverfront – The Interplay of Informality, Flexibility and Exceptionality along the Yamuna in Delhi, India." *Habitat International* 45 (P3): 213–222. https://doi.org/10.1016/j .habitatint.2014.02.007

Frederick, Oliver. 2017. "Neglected Heritage: The inside Truth about Chhota Imambada's 'Beauty'." *Hindustan Times*, March 18. https://www.hindustantime s.com/lucknow/neglected-heritage-the-inside-truth-about-chhota-imambada-s-b eauty/story-GZyRgzVLh2mRXQ3pmhTztK.html

Haider, Mohammad. 2012. *Correspondence Letters and Site Dossier.* Lucknow.

Hay, Sidney. 1939. *Historic Lucknow.* Reprint. New Delhi: Asian Educational Services.

Husain, Yusra. 2017. "ASI Report Comes True, Chhota Imambara 'Restoration' a Failure." *Times of India*, March 18. https://timesofindia.indiatimes.com/city/luck now/asi-report-comes-true-chhota-imambara-restoration-a-failure/articleshow /57697768.cms

Kalshian, Rajesh. 1996. "The Taj Mahal Man." *Outlook.* https://www.outlooki ndia.com/magazine/story/the-taj-mahal-man/201409

Krishna, Ashima. 2014. *The Urban Heritage Management Paradigm: Challenges from Lucknow, An Emerging Indian City.* Cornell University. http://ecommons .library.cornell.edu/handle/1813/36044.

Krishna, Ashima. 2016. "The Catalysts for Urban Conservation in Indian Cities: Economics, Politics, and Public Advocacy in Lucknow." *Journal of the American Planning Association* 4363 (January): 1–4. https://doi.org/10.1080/01944363.20 15.1132390

Krishna, Ashima. Forthcoming. "Politics, the Public, and Urban Conservation in Lucknow, India." In *Untitled*, edited by Bishwapriya Sanyal.

Lalchandani, Neha. 2018. "Sibtainabad Imambara Gets Heritage Bylaws." *Times of India*, December 1. https://timesofindia.indiatimes.com/city/lucknow/sibtainabad -imambara-gets-heritage-bylaws/articleshow/66890227.cms

Llewellyn-Jones, Rosie, and Saleem Kidwai, eds. 2011. *Hazratganj - A Journey Through the Times.* Lucknow: Bennett, Coleman & Co.

Lucknow Bench of Allahabad High Court. 2014. *Syed Mohammad Haider Rizvi (PIL) vs Union of India Thr.Secy.Deptt.Of Culture on 12 May, 2014*. Union of India. https://indiankanoon.org/doc/77706805/

Lucknow Bench of Allahabad High Court – Court No.2. 2013a. "C.M. Application No. Nil of 2013; Maqbara Road and Compound Welfare Society, Lucknow in Response to MISC Bench No. 3173 of 2013."

Lucknow Bench of Allahabad High Court – Court No.2. 2013b. "Misc Bench No. 3173 of 2013; Order Dated November 27, 2013."

Lucknow Bench of Allahabad High Court – Court No.2. 2013c. "Misc Bench No. 3173 of 2013; Order Dated October 10, 2013."

Lucknow Bench of Allahabad High Court – Court No.2. 2013d. "Misc Bench No. 3173 of 2013; Order Dated October 23, 2013."

Lucknow Bench of Allahabad High Court – Court No.2. 2015. "C.M. Application of 2015 in Re. Writ Petition No. 3173 (M/B) of 2013."

Mehrotra, Rahul, and Abha Narain Lambah. 2004. *Conservation after Legislation : Issues for Mumbai*. Mumbai: Urban Design Research Institute.

Pai, Sudha. 2004. "Dalit Question and Political Response: Comparative Study of Uttar Pradesh and Madhya Pradesh." *Economic and Political Weekly* 39 (11): 1141–1150.

Rishi, Meghna, Sanjeev Bhanawat, and Vipul Bajaj. 2012. "Social Media and Its Impact on Cultural Change." *Context: Built, Living and Natural* 9 (2): 97–104.

Roberts, Nancy, and Alasdair Roberts. 2010. "A Great and Revolutionary Law? The First Four Years of India's Right to Information Act." *Public Administration Review* 70 (6): 925–933. https://www.jstor.org/stable/40927109

Shaikh, Juned. 2012. "Who Needs Identity? Dalit Studies and the Politics of Recognition." *India Review* 11 (3): 200–208. https://doi.org/10.1080/147364 89.2012.705636

Stark, Ulrike. 2011. "Associational Culture and Civic Engagement in Colonial Lucknow The Jalsah-e Tahzib." *Indian Economic & Social History Review* 48 (1): 1–33. https://doi.org/10.1177/001946461004800101

Supreme Court of India. 1996a. M.C. Mehta vs Union *of* India (Uoi) And Ors. on 15 March, 1996. Union of India. https://indiankanoon.org/doc/1367896/

Supreme Court of India. 1996b. M.C. Mehta vs Union *of* India and Ors on 30 December, 1996. Union of India. https://indiankanoon.org/doc/1964392/

Supreme Court of India. 1997. M.C. Mehta vs Union *of* India (Uoi) *and* Ors. on 17 March, 1997. Union of India. https://indiankanoon.org/doc/1285836/

Supreme Court of India. 1998. M.C. Mehta vs Union of India (Uoi) and Ors. on 24 March, 1998. Union of India. https://indiankanoon.org/doc/895913/

Supreme Court of India. 2000. M.C. Mehta vs Union *of* India (Uoi) *and* Ors. on 8 November, 2000. Union of India. https://indiankanoon.org/doc/1080796/

Supreme Court of India. 2003a. M.C. Mehta vs Union *of* India (Uoi) *and* Ors. On ... on 18 September, 2003. Union of India. https://indiankanoon.org/doc/1904350/

Supreme Court of India. 2003b. M.C. Mehta vs Union *of* India (Uoi) *and* Ors. on 7 May, 2003. Union of India. https://indiankanoon.org/doc/1746156/

Supreme Court of India. 2006. M.C. Mehta vs Union of India & Ors on 27 November, 2006. Union of India. https://indiankanoon.org/doc/193151/

Supreme Court of India. 2007. M.C. Mehta vs Union *of* India & Ors on 10 October, 2007. Union of India. https://indiankanoon.org/doc/1546599/.

Supreme Court of India. 2010. M.C. Mehta vs Union *of* India & Ors. on 18 January, 2010. Union of India. http://www.advocatekhoj.com/library/judgments/index.php?go=2010/january/72.php

TNN. 2013. "Is the ASI Only a Blue Plaque?" *The Times of India*, August 18. https://timesofindia.indiatimes.com/city/lucknow/Is-the-ASI-only-a-blue-plaque/articleshow/21888146.cms

TNN. 2016. "Heritage Work Stalled: Activists Threaten to File Contempt Plea." *Times of India*, July 26.

TNN. 2018. "2 Heritage Monuments Collapse after Spell of Rain." *Times of India*, August 7. https://timesofindia.indiatimes.com/city/lucknow/2-heritage-monuments-collapse-after-spell-of-rain/articleshow/65300382.cms

5   Heritage education
An essential element in elementary education

Michael A. Tomlan

Introduction

The good news is that an increasing number of people in India at all age levels are reading more. The additional good news is that school enrolment is improving in the country, particularly as food is made available. Even better, educating girls is increasingly recognized as important because this makes them more likely to insist their own children are educated. With a growth rate of 1.25%, India will soon supersede China as the most populous nation. More importantly, the age structure of India's population underscores the need to address early education, as young people are the majority of the population.

The importance of education was recognized in the Indian Constitution, and the 2009 Rights of Children to Free and Compulsory Education Act reaffirms that schooling is free and compulsory for children from 6 to 14 years old. Primary school and middle school students are taught in government schools and private schools, but many of the first group lack adequate facilities, infrastructure, and funding, and they suffer from the limited number of teachers. In stark contrast, the private schools and the international schools in the major metropolitan areas provide hope and suggestions for possible improvements.

Everyone in India understands that the government rules and regulations specify curriculum materials, syllabi, and examinations for the schools. Yet, the high pupil-to-teacher ratio and the lack of well-trained teachers contributes to poor student performance. The challenge remains that the quality of instruction varies very widely, in spite of the widespread emphasis on technological assistance. The dropout rate is far from acceptable. On one hand, India is producing some of the best educated scholars, professionals, and technicians in the world, while on the other hand the general level of education remains low. In this regard, the need to connect a wider group of the public, familiar with the rich variety of rituals, customs, and crafts offers an opportunity to make learning more exciting for young people.

This chapter urges an immediate increase in the amount of time and effort dedicated to examining and interpreting the heritage of India by supplementing the education system. Heritage is at risk throughout the country because it is not widely perceived as an important public good. While math, science, and language are fundamental, it often seems to be a luxury to spend the time and effort to discuss history, geography, and link them all to the historic sites and activities, many of which are nearby. The performing arts and the visual arts provide additional avenues for exploration but, despite the considerable amount of sensitivity to the cultural differences in India, it is comparatively rare to explore the depth, variety, and range of India's heritage in schools.

The early models

The early public education about the heritage of the country arose alongside the "culture of collecting", most often associated with antiquarian collectors, botanists, and mineral explorers. The Indian Museum in Kolkata, established permanently by the Asiatic Society of Bengal in 1814, is the oldest and largest museum and has rare collections of antiques, armour and ornaments, fossils, skeletons, mummies, and paintings (Markham and Hargreaves, 1936, 5). The publications of the Asiatic Society provide ample evidence that similar cabinets of curiosities grew into museum exhibits and, when they would be made available to the public, displayed thousands of items, with sparse labels providing names, dates, and places. In addition, there are numerous tales of travellers during the pre-Colonial and Colonial period describing officials collecting or documenting religious works of art and architecture in their journeys across the country. Noteworthy among these are the contributions of James Fergusson (1806–86), responsible for giving India the first comprehensive works on the history of architectural forms and styles (Fergusson, 1876). His architectural pursuits documented sites with photographs, lithographs, engravings, and drawings, all of which give modern researchers a detailed database on India's architectural monuments linked to the understanding of geology and ethnography (Mullane, 2015, 46–66). More generally, lantern slide lectures sponsored by the Archaeological Survey of India or other societies were important events, but they were few and far between, and often not open to the public. To most Indians, art appreciation, conceived as so much English connoisseurship and antiquarianism and supplemented by increasing historical nomenclature, was a remote activity, conveyed in the specialized language of a foreign country (Tapati Guha-Thakurta, 2004, 43–82).

By 1900, 26 museums were operating in India, while by 1935 the number had grown to 105. Few, however, had any relationship with elementary or secondary schools. The Government Museum in Madras seems to have been a noteworthy exception to the rule, regularly offering tours and classes in a coordinated effort with the secondary school teachers in the area (Markham and Hargreaves, 1936, 19). In general, the authority for education in British India was decentralized and left in the hands of local councils.

In this framework, elites could make a difference and they did so, largely in the private schools in Bombay and Madras. Yet, the overall literacy rates in the early twentieth century hovered under 10 percent, and fewer than 1 in 10 children attended primary school. The possibilities for children to learn anything outside of the basics remained limited (Markham and Hargreaves, 1936, 70, 71).

The paradox of the relationship between England and India is made more obvious when considering the number of pre-Independence thinkers who pointed out the difficulties with British tendencies in education and urged Indians to develop their own system. For example, Sri Aurobindo warned that the roots of Indian culture were being lost; and Rabindranath Tagore noted the widespread dissatisfaction with the foreign standards of comparison that were regularly superimposed, ignoring the contributions of Indians in almost every sphere of education (Danino, 2004, 1).

Contemporary models remain limited and short of the mark

After Independence, the government took a renewed interest in providing free and compulsory education. Basic education, as explained by Mahatma Gandhi, is essentially an education for life, and an education through life. For primary school, craft teaching and productive work habits were taught alongside basic reading, writing, and arithmetic. The links to the academic subjects in the curriculum of "public schools", originally for the sons of gentleman from 8 to 18, grew in the 1950s and 1960s as a greater number of secondary schools evolved with a curriculum of three languages, mathematics, science, history, social studies, and a range of electives (Singh, 1972, 27). Over 80 percent of the students were from urban and metropolitan areas, however, and the institutions charged fees that put education beyond the reach of the average family.

India responded to the call to join the United Nations in 1946, and thus has long compared educational aims and programmes with not only England but a wide range of other countries. A series of conventions and assemblies in India record the continued discussion and slow progress, just as the differences between urban and rural education continued to be evident to observers. In 1962, in collaboration with the United Nations Educational, Scientific, and Cultural Organization (UNESCO), the Institute of Education Planning and Administration was established with the broad goals of training administrators, teachers, and representatives of state and local governments. Renamed the National Institute in 1979, the emphasis on teacher training and need for the ever-changing re-education of local officials continues.

The link to the world's heritage, and the recent role of museums in educating the public is easily described in recently produced publications. Educational materials to advance primary education have been produced by UNESCO and made available in India since at least 1978 ("UNESCO

Publications on Primary [sic.] Education (1978–1991)", 1992, 285). The idea of the world's heritage was first developed in Europe and made available to Indian educators who expressed interest in the curricula and lesson plans. Similar to the lessons taught in museums, the UNESCO linkages provide "Heritage in Young Hands" kits, designed "to introduce students to the basic research methods, searching for and analysing information, drawing conclusions and formulating suggestions for action with respect to world heritage conservation" in each country, region, or specific location. The materials assume that the schools will have access to the statistics, information, research findings, and support material either by the internet or by local libraries. The kit, available in 12 languages, allowed the materials to be used in classrooms worldwide. In 1981, when UNESCO declared April 18th as World Heritage Day and asked all signatories to the World Heritage Convention to involve local communities in the celebrations, children were seen as essential participants and the idea of using the kits spread (Agrawal, 2007, 449). In India, the Department of Culture introduced the idea of a Heritage Week in 1986, and the country began to celebrate World Heritage Day in the same fashion, enlisting school children. In this regard, the efforts of the Indian Trust for Art and Cultural Heritage (INTACH) are laudable. By approaching the schools in Delhi, Chennai, Hyderabad, and Ahmedabad with the idea of establishing young explorers through "heritage clubs" and encouraging the students to "adopt" natural and built heritage sites and properties, a variety of local initiatives are being tested. At the same time the distribution of these tools demonstrate they are only the first step for preparing regional and national activities.

A number of private schools have made remarkable inroads in urban areas by testing and improving these heritage activities. Whether rural or urban, community-based projects seem to be having the best results. Working with villages and local schools, for example, the "Read India" Program has made a significant impact outside of the school system's limitations by enlisting non-governmental actors and providing, repeatedly through the year, "learning camps" for children for 6 to 10 days at a time (Banerji and Chavan, 2016, 453–475). This suggests that by moving off in another direction, change in the school system can occur, and this is where heritage education can play an essential role. Fortunately, after putting aside their textbooks, children have an increasing number of books that touch on the traditions and myths of what might be considered "intangible cultural heritage". For example, the Amar Chitra Katha titles, introduced in 1967, number in the hundreds and are widely available in over 1,000 retail outlets. They include the "Heroes of Hampi" series shown in Figure 5.1.

Another approach

By contrast to the foregoing NGOs, the Indian Traditions and Heritage Society (ITIHAAS) approaches the need for heritage education in another

Figure 5.1 Of the available illustrated children's books about historic sites, one
of the more popular is the imaginative "Heroes of Hampi". Source:
Author, 2016

fashion. The founding director, Smita Vats, wondered how her daughter
would learn the stories that were so important to her as a child. She engaged
six schools the first year, drawing on her decades of experience in theatre
and film with storytelling on children's walks. Ms. Vats documented the
oral histories of more than 200 residents of the walled city of Delhi to form
stories used to stimulate the imagination of the children. Using these initiatives to stimulate others, the organization she founded in 2004 has grown to
include children in 700 middle and senior schools in the region, with music
and dance. Meanwhile the network has expanded and brought together policy makers, school principals, teachers, and corporate sponsors to discuss

intangible and tangible heritage in annual meetings and smaller groups. Ms. Vats has cast a distinctly proactive tone from the Delhi base of the organization, and has moved urban pupils into rural settings. ("ITIHAAS – Indian Traditions & Heritage Society – YouTube", July 9, 2010, accessed November 20, 2017).

A typical discussion

The intellectual distance between the recognized importance of the historic area and the ongoing activities of education at the primary and secondary education in the community is often all too evident ("Karnataka Rural Analysis Based on Data from Households", 2016, 125–129). The Government Secondary School in Hampi provides a typical case. Established in 1958, it is operating to the 8th standard, enrolling about 250 students with 8 teachers. As in many rural schools in Karnataka, the limitations of the personnel and the facilities become apparent when attempting to introduce new initiatives. In discussions of a pilot heritage project with the Village Panchayat, the principal and the teachers, it quickly became clear that introducing any discussion about the significance of the region to the students would be difficult. Village leaders saw that any funding for the effort would be chiefly directed to the students. Meanwhile the principal worried that the teachers would have difficulty mastering any of additional material and working it into the classes, particularly in the time allotted in the curriculum. The students were already woefully behind in their required coursework. The teachers also worried about when they would have the time to prepare for class, because they had time only on the weekend, and that was dedicated to their family obligations – washing, sewing, ironing, renewing the bedding and preparing food for the next week. In short, the work at home demanded their full attention, sometimes even when they should be teaching at school. An additional issue arose when it was proposed that the students might visit a local museum because considerable "red tape" was involved, with permission from the parents, the availability of the site staff, guards, and transportation. Last, but by no means least, the need to provide transportation, lunch, water, and snacks created additional hurdles in establishing such a programme (Figure 5.2).

Each of these problems are addressed in turn. To begin to teach the teachers, discussions about the overall curriculum, and the length of the training programmes with interactive sessions had to be considered and organized. Local university faculty and staff were recruited to assist with the specifics of translation. Workshops, heritage walks, and other exercise modules were designed by the teachers, along with evaluations. Because the teachers had only weekends to dedicate to learning new subject matter, the early classroom trials were experiments, as they required more interactive student participation, showing how learning could be fun and enjoyable. Student prizes and incentives were added.

Figure 5.1 Lessons and outdoor performances for the children attending Hampi's school provide an opportunity to display their knowledge of local, regional, and national culture. Source: Author, 2014

Meanwhile, in some cases the teachers had to procure weekend babysitters: an added expense. Looking ahead, theoretically any material produced for the Hampi school could almost immediately be improved in another iteration. This suggested introducing material on a "trial" or a "pilot" basis to be able to adjust before reproducing it, for as any good teacher understands, mastering the new information takes more than a year and a single group of students. Hampi became the theoretical starting point, moving on to nearby schools in Kaddirampura, Kamalapura, and Anegundi (Mooij, 2008, 508–523).

The need for recommitment

Working with the educators in existing schools, more partnerships with heritage organizations and museums are needed. Heritage education should be introduced at every opportunity, examined from the point of view of each subject in the curriculum. Supplying schools with services, including curriculum materials and speakers in the classroom and on tours is a good first step. As has been seen in other countries, involving teachers in the development of ongoing programmes is even better. This strengthens the value of well-educated teachers to school administrators and local officials, who in turn can share the accomplishments of the students with the public. It should be recognized that the teachers will continue to feel the pressure of getting their students to pass standardized examinations, a crucial feature in their evaluations on a number of levels. That should not be the only measure of the students' educational success, however. By encouraging the students to make observations, take initiatives, and imaginatively use first-hand evaluations has the potential to enhance and improve the delivery of universal elementary education (Khastagir 2016, 91–105). Help is needed, as an increase in government spending is not ever likely to provide the amount of assistance for the growing number of youth in the nation. Not only the parents and teachers should become more involved with heritage education, but professionals in all disciplines should help. Developing an increased sensitivity in students to culture today cannot help but make their voices heard in the decades to come.

References

ACER Center. 2016. "Karnataka Rural Analysis Based on Data from Households." *Annual Status of Education Report*, 125–129. New Delhi: ACER Center. http://img.asercentre.org/docs/Publications/ASER%20Reports/ASER%202016/aser_2016.pdf

Agrawal, R. C. 2007. "Heritage Education: A Case Study in India," in *Heritage and Development. Recent Perspectives*. Papers Presented at the 12th International Conference of National Trusts, New Delhi, 3rd–5th December, 2012, New Delhi, INTACH, 2012, 448–452.

Banerji, Rukmini and Chavan, Madhar. 2016. "Improving Literacy and Math Instruction at Scale in India's Primary Schools: The Case of Pratham's Read India Program," *Journal of Educational Change*, 17: 453–475.

Danino, Michel. 2004. "Integrating India's Heritage in Indian Education," Paper Presented at the National Symposium on Cultural Education in 21st Century India, organized by Amrita Vishwa Vidyapeetham, Ettimadai, Coimbatore, 8–9 May.

Fergusson, John. 1876. History of Indian and Eastern Architecture. London: J. Murray.

Guha-Thakurta, Tapati. 2004. *Monuments, Objects, Histories: Institutions of Art in Colonial and Post-Colonial India*. New York: Columbia University Press.

ITIHAAS—Indian Traditions & Heritage Society—YouTube, 9 July, 2010. Accessed 20 November 2017.

Khastagir, Saheli. 2016. "Universal Elementary Education in India: A Reality Check," *Social Change*, 42(1): 91–105.

Markham, Sydney Frank and Hargreaves, H. 1936. *The Museums of India*. London: Museums Association.

Mooij, Jos. 2008. "Primary Education, Teachers' Professionalism and Social Class about Motivation and Demotivation of Government School Teachers in India," *International Journal of Educational Development*, 28: 508–523.

Mullane, Matthew. 2015. "The Architectural Fossil: James Fergusson, Geology, and World History," *Architectural Theory Review*, 20(1): 46–66.

Singh, R. P. 1972. *The Indian Public School*. New Delhi: Sterling Publishers.

UNESCO Publications on Primay (sic) Education (1978–1991). 1992. Documents on Primary Education in India (1792 to 1992). J. C. Aggarwal, ed. Delhi: Doaba House.

Part II
Critical challenges in heritage conservation

6 History, memory, and contestation

Challenges in preserving Amritsar's diverse heritage

Gurmeet S. Rai and Churnjeet Mahn

Introduction

The city of Amritsar in the Indian state of Punjab was founded by the fourth Sikh Guru, Guru Ram Das, in 1577 and since then has become one of the holiest and most recognised sites associated with the Sikh faith. The surrounding area was home to other sacred sites, such as the Ram Tirath, and contemporary Amritsar houses various religious monuments, illustrating the diversity of worship in the region. The narrative of Amritsar's significance, however, has been dominated by historical events in the 20th century, namely the massacre of a congregation in Jallianwala Bagh in 1919, violence in the aftermath of Partition in 1947, and the attack on the Golden Temple known as Operation Blue Star in 1984. Jallianwala Bagh and the Golden Temple have remained at the centre of the city's self-presentation of its heritage to tourists, the majority of whom come specifically to visit the temple.

In the past decade several initiatives have come together to investigate how to diversify and enhance the presentation and preservation of Amritsar's rich history. In 2008, for example, after much persuasion by the state government, Gobindgarh Fort was handed over by the Prime Minister of India to the civil government for its protection and conservation. In 2016, the Arts and Cultural Heritage Trust set up what has been called the world's first Partition museum which is housed in the conserved Town Hall. And most significantly, in 2015, the Government of India set up the National Heritage City Development and Augmentation Yojana (HRIDAY), a scheme which identified 12 historic Indian cities, including Amritsar, for investment. This investment related to how inclusive practices of heritage conservation could be developed in conjunction with sympathetic urban infrastructure development for historic sites. The authors collaborated in Punjab to experiment with developing inclusive heritage practice. This chapter offers a snapshot of that work, designed to investigate how to bring 'forgotten' histories, people, and legacies into mainstream heritage narratives.

Challenges at Amritsar

As a city, Amritsar has found itself at a series of crossroads. The artisans and craft guilds brought to the city by Guru Ram Das in turn brought with them a diversity of skills, faiths, and cultures. The community actively participated in the building of the Golden Temple as the followers of Guru Arjan Dev, the fifth Sikh Guru. The site has been evolving and adapting to its socio-political, religious, and cultural context since that time. Through the colonial period, Amritsar continued to be a site of syncretic religious practice and pluralistic approaches to community and identity.[1] This vision of a pluralistic city was most seriously challenged by the Partition of India into India and Pakistan in 1947, marking the end of British colonial rule. Amritsar was at the frontlines of religious and political violence due primarily to its proximity to the new national border. The religious divide caused by partition has had lasting impacts on the city, especially in terms of localised articulations of cultural heritage, and memorialisation (due to the displacement of communities with historical ties to the city).

Further, the events of 1984 at the Golden Temple have variously been described in the language of terrorism, massacre, and genocide. However, they may be interpreted or framed, the events clearly demonstrate the increasingly tense relationship between faith and nation, which has helped reorient Amritsar's history as synonymous with Sikhism at the cost of a 'messier' or more complex narrative. As the home of the Golden Temple, Amritsar has become a spiritual centre for Sikhs across the world thereby making it a site impacted by both pilgrimage and tourism.

The most recent recognition of this has been the significant investment by the Government of Punjab in beautifying the main road leading to the temple and creating a 'Heritage Street' to enhance the experience of pilgrims and tourists, a project completed in 2016. The Heritage Street in Amritsar is a 500m stretch connecting the Golden Temple with the Town Hall and attempts to showcase aspects of Punjabi heritage through food and folk culture, alongside selected national symbols. Life-size statues of bhangra dancers sit near grand statues of Maharaja Ranjit Singh and Dr B R Ambedkar. Over 170 shops have been redeveloped along the street with standardised frontages and improved urban infrastructure. Uniform architecture has been an underlining imperative for the project, which received strong support and funding from the then chief minister of Punjab. The alignment of Punjabi culture, the Sikh Empire and the Indian secular state appears seamless but carries with it a clear message: Punjabi heritage can and should be read in the context of the Indian nation state. Heritage Street is an initiative which has drawn significant criticism for its simplistic presentation of heritage that draws its architectural inspiration from Jaipur rather than Punjab, making 'uniformity' another word for culturally insensitive and inappropriate standardisation (Dutt 2016). The selection of shops which line Heritage Street self-consciously display 'authentic' Punjabi culture, from food to wares primarily for tourists; the newly fashioned facades of Heritage Street mask the traditional markets surrounding the temple.

To step outside of Heritage Street is to re-enter the traditional bazaar of the city which now sits in stark contrast to the homogenised architectural aesthetics of the street with large promenades, visually connected to the Golden Temple through prominent LCD screens projecting a live-stream from the temple. Beyond the Heritage Street, Hindu shrines and temples co-exist with historical mosques and an active bazaar that features trades, crafts, skills, and wares demonstrating Amritsar's position as a site of particularly deep cultural and religious confluence.[2]

The Heritage Street has struggled to offer a model of heritage conservation that is sympathetic to its context, however it demonstrates an increasing desire from government agencies to integrate everyday heritage with city planning – a step in the right direction, which offer opportunities to expand the current heritage management practices in India which have primarily been 'monument-centric'. The collaboration between the authors under the project 'Creative Interruptions' attempted at imagining entirely different processes and outcomes for the Heritage Street as an alternative to the existing state-sanctioned representation and development around the Golden Temple. The following discussion traces part of the project's work through two sites: Rambagh Gate and Rambagh Garden that are connected to the Heritage Street and the Golden Temple, but are currently excluded from the Golden Temple's self-presentation of its history and heritage. The project proposes an inclusive approach to heritage management that engages communities to build capacity around conservation areas and values local knowledge as vital and a form of intangible heritage.

Case study: Rambagh Garden

As the Cultural Resource Conservation Initiative (CRCI) developed the City HRIDAY plan for Amritsar based on cultural heritage mapping, five heritage zones were identified for improvement of infrastructure: physical, social, and/or institutional. The CRCI identified several projects which contributed to the enhancement of both natural and cultural heritage while ensuring that these sites and buildings would contribute to improved quality of life for its inhabitants. A key example of this was the conservation of the Rambagh Gate. When Maharaja Ranjit Singh resided in the Summer Palace in the garden, his visit to the Golden Temple involved a procession through the Rambagh Gate. There was an unbroken line of sight between these two landmarks, which is now completely obscured by the contemporary growth of the city (including a railway line in the colonial period and a contemporary flyover). Metaphorically, the new spatial juxtaposition of old and new demonstrates the ways in which lines of real and imagined connection and mobility across Amritsar have been disconnected, circumvented, and re-routed.

The Rambagh Garden is an 80-acre garden complex organised into four quads and centred around the Summer Palace of Maharaja Ranjit Singh. Built in the early 19th century, the garden served the dual role as the capital seat of governance and a residence in proximity to the Golden Temple. Structures at the intersections of the quads served different purposes from entertainment areas to residence, with the Summer Palace as a central structure (see Figures 6.1 and 6.2). The garden did not exist in isolation: a road was designed to lead directly to the Golden Temple which passed through the Rambagh Gate, which formed part of the wall around the city of Amritsar. Therefore, the structure cannot be read in isolation, it is part of the 19th century city planning and geometry of Amritsar. Unfortunately, the original layout of Rambagh Garden and its historical importance have been steadily eroded. Taxi drivers and locals know the site as 'Company Bagh' from the time of the East India Company when various offices and members' clubs were added to the gardens. The garden is currently protected by the Archaeological Survey of India (ASI) and offers a rare open space in a city struggling under the pressures of rapid urbanisation.

Walkers have access to portions of the site although they remain unaware of its importance in Sikh and Punjabi history and culture. The Rambagh

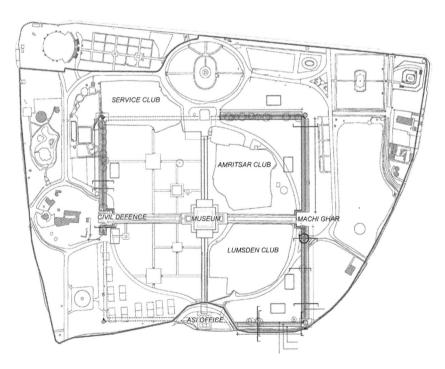

Figure 6.1 Conservation map of Rambagh Garden, 2017, Cultural Resource Conservation Initiative, New Delhi.

Figure 6.2 Summer Palace under conservation, Rambagh garden, 2017, Cultural Resource Conservation Initiative, New Delhi.

Garden in its original iteration did not last long; with the commencement of colonial rule in 1858, the 'open' site of the garden was used by the British to build three clubs: Amritsar, Service, and Lumsden Clubs. Vehicular roads were added within the complex for ease of movement across two busy roads, the four structures across the east-west and north-south axes were given new uses by the British; the eastern gate housed a municipal library and an aquarium (which gave it the name of *machi-ghar*), the western structure (that housed the *darbar* of Maharaja Ranjit Singh) was altered extensively to house the office of civil defence. The northern structure was leased to the service club for their administrative offices. The southern gate came to be protected by the ASI much earlier than the entire garden site, which received ASI protection in 2004. The colonial-era clubs were built in two of the quads while the third is located on the northern edge (partially within the quadripartite garden and partially outside). The walls of the Rambagh Garden were demolished during the British period, however the corner pavilions on raised bastions were retained, hence the form of the garden palace can be determined by their presence. The lease period of the clubs expired in the late 20th century and several civil society bodies (most prominently the Amritsar Vikas Manch) made an appeal to the Punjab and Haryana High Court for the removal of the clubs from the garden precinct. The clubs have now lain their own claim to the site, with their management arguing that they have become part of its historical fabric. The act of conserving the Rambagh Garden offers a case study for the larger issues in conservation in Punjab. It forces a consideration of how new kinds of dynamic processes and dialogues are required to address the challenges of what counts as 'heritage' in Punjab and who its intended beneficiaries are.

Conservation challenges at Rambagh Gate and Garden

The Rambagh Gate (see Figures 6.3 and 6.4) is the only surviving 19th-century gateway built by Maharaja Ranjit Singh (the other 11 were replaced during the colonial period). The police station located in the gate was vacated in 2007 to allow for conservation work and tourism related use. The CRCI undertook the project; its first phase was funded by the Ministry of Tourism, Government of India and implemented by the Department of Tourism, Government of Punjab. The project involved removal of the incompatible material additions and elements of the structure to restore its architectural and structural integrity. In its current state, the Rambagh Gate is surrounded by a bazaar (temporary shops built by Municipal Corporation of Amritsar and given on lease to local residents, many of whom relocated to Amritsar at the time of the Partition of India in 1947). On the rampart there is a printing press that has been in operation since the early 20th century, alongside a government school. The structures housing the two uses of the

Figure 6.3 Rambagh Gate under conservation, 2017, Cultural Resource Conservation Initiative, New Delhi.

Figure 6.4 Rambagh Gate completion of conservation work, 2018. Cultural Resource Conservation Initiative, New Delhi.

printing press and the school were built in the early 20th century (with building materials obtained from the earlier period buildings).

The questions related to heritage and cultural narratives of the community were as follows: should the conservation of the gate and rampart involve removal of the colonial period additions and uses, which would include the printing press and the school? Does 'good conservation practice' of the gate mean 'monumentalising' the structure, thereby creating a 'dead edifice', even if it is not feasible to activate the gate structure as a working thoroughfare. Should urban renewal in the setting of the gate, which is part of the heritage zone of HRIDAY, involve removal and relocation of the shops and the bazaars? The perception and association of the local community with the gate structure has been an ongoing question and negotiation – do they see the gate as 'their heritage' worthy of conservation?

This conservation project for the Rambagh Gate was revived as part of the HRIDAY programme. This phase of the project addressed the completion of the conservation of the gate and the rampart along with the printing press and the school. The project team, comprising of authors, oral historians, and conservation architects in consultation with the workers in the bazaar adjoining the gate, determined how conservation could take place without long-term displacement or disruption of the school, printing press, or bazaar. The first stage of this work involved a team conducting a series of interviews with people working in the immediate vicinity of the Rambagh Gate, where there is a small bazaar. Along with interviews, cultural mapping was conducted to map the types of wares sold across the bazaar, and document the range of existing skills and crafts in the area. Consultations were undertaken to understand the association of the community with the gate. All this information helped to generate the adaptive reuse brief for the gate and propose some specially commissioned interventions into the existing arts and crafts in the Rambagh Gate area to connect the site back to the cultural fabric of the city.

The exhibits within the gate tell the story of the cultural history of the city with a special focus on vernacular creativity, which is reflected in the title of the museum: Lok Virsa (Museum for the People's Culture). A cultural/recreational zone around the Rambagh Garden and the gate was envisaged from where it could be networked with sites such as the Thakur Singh Art Gallery, the Shahidi Bor (the banyan tree associated with the Kuka Movement), the local foods and crafts complex at the refurbished Guru Tegh Bahadur Hospital, Rambagh Police Station Interpretation Centre (to be housed in the conserved historic wall of the city), Christ Church Cathedral and others sites of local history, culture, and living traditions.

Conservation interventions in the Rambagh Garden required the removal of the vehicular movement to recover the garden as a pedestrian space (with parking lots on the periphery). Investigations were undertaken in the garden to determine the historic layout of the quadripartite garden after obtaining approval from the ASI. The original water channels recovered through excavations were restored and new pathways were laid in two of the quarters to revive the 'spirit of the garden'. This was possible on one half of the garden enclosure as the other half is occupied by the clubs. The building of the Summer Palace, structures on the east-west axis and the corner pavilions were restored. As part of the HRIDAY project, 'stub walls' (low walls) with pathways are in the process of being built on the footprint of the fortification enclosure of the palace garden (demolished by the British). Efforts are being made to recover the garden palace as a historical edifice to Maharaja Ranjit Singh as the site owes its significance to this historical narrative. These have been the first few steps to begin the conversation between the various stakeholders and community about this site. The stakeholders include the Municipal Corporation of Amritsar (the owners of the property), the ASI (as the site is protected under the Ancient Monuments and Archaeological Sites and Remains Act, amended and validated 2010), the Department of Cultural Affairs, Archaeology, and Museums (Government of Punjab, who manage the Maharaja Ranjit Singh Museum housed in the Summer Palace), as well as the three private clubs.

In the recent past, the Amritsar Improvement Trust has made several interventions in Rambagh Garden by way of large car parks and an outdoor exhibition of stuffed animals on the site without seeking any approvals or consultations from the ASI, thus compromising the integrity of the garden. On the other hand, a project undertaken under the HRIDAY program, with the permission of the ASI, to define the edges of the historic garden introduced stub walls on the original footprint of the walled enclosure with a 5.0 metre-wide pathway along these walls, so as to recall the memory of the fortified quadripartite garden complex.

In the absence of a shared vision and a conservation management plan for the site which recognises the elements of value and needs of the community, the interventions of the Trust are unsympathetic to all that is of value on this site. There remains the need for a shared vision that can be facilitated through a platform for dialogue but who should lead this? Would a legal framework better enable this effort? How should it be imagined in the absence of the municipal authorities and the ASI interest in participatory processes?

The importance of history, memory, and heritage

There has been a significant body of writing on forms of cultural amnesia, memory and memorialisation in Punjab, especially around partition.[3] Increasingly, studies have focused on the reverberations of partition that continue into the present, from government policies to religious and cultural tensions with the nation state (Kaul 2001; Zamindar 2007; Mahn and Murphy 2017). However, it is also important to remember the ways in which pre-partition history can be selectively deployed to align with the interests of the present, namely through foregrounding a Sikh narrative. Not engaging the story of a pluralistic Amritsar is a choice rather than a reflection of historical evidence of settlement. The role of colonial history here is also significant, and the remnants of colonial architecture in the city's lesser-known and unrecognised industrial (and institutional) heritage for instance in the former powerhouse on the Upper Bari Doab Canal, and the pump house in the park known as the 40 *khuh* (40 wells), offers an opportunity to identify different contextual frames to presenting the history of British intervention alongside sites such as the private members' clubs in the Rambagh Garden.[4]

As part of a recent project, conservation architects from CRCI conducted surveys and interviews around several heritage sites in Amritsar including the Rambagh Garden and the Rambagh Gate. Surveys around the Rambagh Gate were undertaken to understand the way the sellers in the bazaar adjoining the site perceive the gate and its history. Talking about and discussing heritage was designed as a process of inclusion, but also of learning for the CRCI team. The following excerpt is from a CRCI report on this survey in relation to the history of Amritsar around the gate which responds to the question: how do the shopkeepers around the gate view this slice of history in their midst?

> People have various levels of historical understanding [of Amritsar], and all of them agree on three points – that the Gate was made by Maharaja Ranjit Singh, that the bazaar grew during British times, and also that the area was predominantly a Muslim area before the Partition, and

drawing from this point [...] On the matter of history and memory, I am of the opinion that more strong and rooted traditions don't exist and the connection to history is almost secondary because the inhabitants of the place don't trace a lineage of more than three generations at this place because of the Partition.

(Cultural Mapping in Amritsar 2017)

The different routes to historical veracity (from empirical studies to forms of vernacular truth) compete in defining the space and its historic association and significance. The lack of lineage in the extract is complex to read: as a microcosm for diversity in northwest India, Amritsar's history offers multiple points of entry into cultural belonging. To say its history is heterogeneous is hardly controversial, what is controversial, however is the increasing homogenisation of Amritsar's historic interpretation through heritage sites. What are the challenges and rewards of opening heritage and conservation practices to more inclusive conversations, especially when those may be at odds with competing historical 'truths'? In their discussion of community inclusion in heritage, Emma Waterton and Laurajane Smith offer an overview of how radically different perspectives have the potential to produce incompatible views of conservation:

Not only are many people overlooked as authorities capable of adjudicating their own sense of heritage, so too is their lack of access to necessary resources. They are, in effect, subordinated and impeded because they do not hold the title 'heritage expert', and are lacking in resources assumed necessary to participate in heritage projects (Western schooling, economic means, etc.), and also potentially 'lacking' a particular vision or understanding of heritage and the accepted values that underpin this vision (universality, national and aesthetic values, etc.)

(Waterton and Smith 2010, 10)

The CRCI's philosophy and practice has been informed by traditionally marginalised or ignored voices in heritage management. In Punjab it is not remarkable to see Dalit Sikhs taking part in Christmas processions, Sikhs going to *dargahs*, or Christians looking after *dargahs*. There is a gap between the complexity of lived practices and the discourses people adopt to match their own identity. An inclusive practice of heritage management in this context involves a reciprocal process of learning where an area undergoing conservation does not harm or destroy an organic and vital community life around it, in this case the bazaar surrounding the gate. Especially in this case, safeguarding forms of intangible heritage (including speciality foods and goods) alongside tangible heritage (the gate) is a way of ensuring the site remains relevant and open to the community currently associated with it.

One of the processes CRCI has engaged in is allowing heritage narratives and opinions to be expressed without immediate judgement. This, necessarily, introduces new agendas and imperatives. It would be wrong to believe that community voices are automatically part of the authentic heritage of a place, but it does not mean that the economic or social and cultural aspirations of those voices are not authentic. For example, when questioned about the possibility of turning the conserved Rambagh Gate into a People's Museum, the shopkeepers of the bazaar had a variety of opinions.

On the matter of cultural representation, all people agreed to the idea of the museum. They believed that a museum will profit them all. On being told about the museum, they imagined a space like Heritage Street, which according to them will attract a lot of tourists and will be a boost to the businesses. They also believed that the area will then become 'cleaner' and more 'ordered'. Some of the shopkeepers expected that the municipality shops around Rambagh Gate have to be removed. But when it came to contents of the museum, the people were divided on whether they want it to represent and glorify Ranjit Singh, or whether they want it to document the history of the bazaar ('Cultural Mapping in Amritsar' 2017).

The shopkeepers around the Rambagh Gate were keenly aware of the financial opportunities presented through the commodification of heritage, but the mixed opinions around presenting the story of Maharaja Ranjit Singh and offering a history of the bazaar offers routes into seeing, presenting, and displaying Amritsar's history in ways that connect across its historical spectrum including stories of occupation, migration, and a localised sense of belonging in the post-partition decades. The city of Amritsar offers the material evidence to share various stories about *Punjabiyat*.[5] However, finding the balance in their articulation and giving space to diversity demanded the project include the missing voices of women's groups and younger generations.

Conclusion: democratising conservation practices in Amritsar

The challenge in recovering the Rambagh Garden demands creative dialogues between different stakeholders including communities and organisations associated with it. However, to re-situate the garden more meaningfully in the fabric of Amritsar involves imagining a reconnection to the Rambagh Gate from the Golden Temple to clarify the historical significance of the heritage sites and the route connecting them. The reimagining of this historical link and route into Amritsar, if only through historical consciousness, will help bring these sites into a new relief in the city. To avoid the narrow aesthetic vision of Heritage Street which is not sympathetic to local architectural style or cultural influences, the involvement of local shops, craftspeople, and communities needs to be at the heart of consultation rather than physically and culturally marginalised by the forces of an imagined aesthetic for 'heritage tourists'. It is, however, difficult to integrate the diversity of these voices so that the conversation is not dominated by the short-term financial interest of the minority.

Heritage can be viewed as an enterprise, but its value is impossible to quantify. Identifying and including a broader range of voices and ensuring representation across genders, castes, and faiths is an important next step for all work in heritage practice in Amritsar. The challenge of this kind of intervention is that it must balance the economic opportunities generated through projects like the Heritage Street with a conservation approach which is sympathetic to local architectural typologies and can present diverse narratives in a city. The current work at the Rambagh Garden and Rambagh Gate does not provide easy solutions to the challenge of creating truly inclusive practices of heritage and conservation management. However, it does demand an interruption to the state's disproportionate investment in a few sites for religious, ideological, and economic imperatives, which strategically disinvests in, and devalues, the rest of Amristar's heritage.

Notes

1 A growing body of work has demonstrated the complexity of cultural and religious practice in Punjab during the history of Sikhism. The instigation of the Colonial State and subsequent nationalism has worked to harden the borders between identities, especially around issues of caste and religious identity. For a comprehensive discussion of how Punjabi as a language became subjected to this linguistic politics, see Mir (2010).
2 Key historical examples include the Durgiana Temple and the Khairuddin Mosque.
3 For an overview of how cultural amnesia can be understood as a cultural phenomenon in this context see, Kabir (2013).
4 Sikhism's close historical connections to the faiths in the northwest of the Indian subcontinent is particularly important as it demonstrates key points of cultural and philosophical interchange. See Singh (2004).
5 The *Punjabiyat* can be variously read as a political project based on a Punjabi nationalism that sidesteps Sikh nationalism, and a utopian project to read Punjabi identity across the Indo-Pak border. See Singh (2014).

References

"Cultural Mapping in Amritsar". 2017. *Cultural Resource Conservation Initiative.* New Delhi: Cultural Resource Conservation Initiative.

Dutt, Nirupama. 2016. "Amritsar's Makeover: Golden Grandeur with a Heritage Tinge." *Hindustan Times*, 24 October. http://www.hindustantimes.com/punjab/ht-special-amritsar-gets-a-majestic-makeover-golden-grandeur-with-a-heritage-tinge/story-0GisnbT7dbOtJj4l6fG2aI.html.

Kabir, Ananya Jahanara. 2013. *Partition's Post-Amnesias: 1947, 1971 and Modern South Asia.* New Delhi: Women Unlimited.

Kaul, Suvir, Eds. 2001. *The Partitions of Memory.* Bloomington, IN: Indiana University Press.

Mahn, Churnjeet and Anne Murphy, Eds. 2017. *Partition and the Practice of Memory.* London: Palgrave.

Mir, Farina. 2010. *The Social Space of Language: Vernacular Culture in British Colonial Punjab*. Berkeley, CA: University of California Press.

Singh, Khushwant. 2004. *A History of the Sikhs: 1469–1838*. Oxford: Oxford University Press.

Singh, Pritam. 2014. "Class, National and Religion: Changing nature of Akali Dal politics in Punjab, India." *Commenwealth and Comparative Politics* 52 (1): 55–77.

Waterton, Emma and Laurajane Smith. 2010. "The Recognition and Misrecognition of Community Heritage." *International Journal of Heritage Management* 16 (1–2): 4–15.

Zamindar, Vazira Fazila-Yacoobali. 2007. *The Long Partition and the Making of Modern South Asia: Refugees, Boundaries, Histories*. New York: Columbia University Press.

7 Loss of cultural artefacts

Continuing challenges around antiquities trafficking from India

Swapna Kothari

Introduction

The cultural wealth of India encompasses a rich and diverse set of antiquities. Unfortunately, many sites and objects comprising this cultural wealth remain undocumented and unprotected, vulnerable to pillaging and looting. The illegal antiquities trade is therefore a matter of immediate concern within the cultural heritage discourse in India. Looting[1] and trafficking of Indian artefacts have had centuries-long history, and continue to be mostly unaddressed. This chapter highlights recent issues and developments in the Indian antiquity trade and relevance of antiquity trafficking in the heritage conservation field. The illegal procurement and movement of these artefacts[2] is facilitated by ineffective laws and limited infrastructure for their protection. Recent thefts brought to light by international media are discussed in later sections, providing an insight into the operation of the illicit artefact trade network in India. The concluding paragraphs describe possible strategies that could address the identified problems. In light of the global repatriation efforts, it is important to discuss the antiquities trade and its relevance to the conservation field.

A brief history of antiquities conservation in India

India's national-level cultural heritage protection, principally carried out by the Archaeological Survey of India (ASI) under the Ministry of Culture, maintains built structures and sites of national importance and advances archaeological activities and research. The ASI's Antiquity section is entrusted with implementing legislation, regulating the export of antiquities, and also preventing their illicit trade. One of its primary divisions, the Central Antiquity Collection (CAC), established in 1910, collects and curates excavated objects from archaeological sites across India, either at on-site museums or via off-site storage. The CAC also administers the repatriation of stolen antiquities, receiving them at its centre, in Purana Qila, New Delhi. If the antiquities have been registered as stolen, they are repatriated, once they are sent to investigative authorities. In cases lacking claimants, the antiquities are sent to museums near their place of origin. The CAC also has a Data Bank, which is a subsidiary of the ASI's Antiquity Section and located in the Red Fort complex. It was established in 1976 for the safekeeping of documents related to the certificates of registered antiquities.[3] These documents, received from various registering officers of the state archaeological departments and the ASI circle offices, relate mainly to the transfer of ownership. In 2007 India's Ministry of Culture launched the National Mission on Monuments and Antiquities (NMMA) to address the documentation of what officials estimate are more than seven million excavated antiquities. The NMMA had a five-year mandate to create a national database for the antiquities.[4] As of 2018, however, just over 700,000 antiquities were listed on its website, and the work is not proceeding any faster (NMMA 2018, 1).

Even as the documentation continues slowly, the loss of cultural artefacts continues at a faster pace. The Ministry of Culture recorded 101 antiquities stolen during 2000–2016 from protected sites, while the National Crime Records' data reports 4,115 cases of stolen 'cultural property' during a span of just four years from 2010–2014 (Gupta 2016, 1). The huge discrepancies in numbers are unaccounted for and have not been clarified by any official agency.[5] Artefacts that are stolen range from miniature and massive stone statues, to ivory and wood carvings, lamps, figurines, prayer (*puja*) utensils, maps, church altars, jewellery, as well as manuscripts and paintings. They disappear from unprotected and remote sites as well as from nationally listed ones. The 12th century sculpture of Brahma, stolen in 2001 from Rani-ki-vav in Patan, a World Heritage Site, shows that even protected sites are at risk.[6] Although it was found after being missing for 15 years at an exhibition-cum-sale gallery in London[7] (Indian High Commission 2016, 1), it only made its way to Delhi by 2018. This raises questions about the effectiveness of the legal and administrative apparatus in the timely repatriation of antiquities.

Current legislations for antiquity protection

Explicit regulations on export of antiquities[8] were developed after India achieved independence in 1947. The Antiquities Export Control Act of 1947 was introduced with a view of making 'better provision for controlling the export of objects of antiquarian or historical interest or significance' (Thapar 1984, 66). The export of antiquities required a license granted by the Government of India. This act was superseded in 1972 by the one currently in effect: the Antiquities and Art Treasures Act (AAT). This improved act among other things further regulated the export of antiquities to prevent smuggling and other fraudulent dealings. The AAT also mandated all objects be registered with the ASI[9] and empowers the central government to confiscate antiques from their owners with paid compensation based on an arbitrary market assessment[10] (Archaeological Survey of India 2017). The

Expert Advisory Committee in ASI's 29 circles issue non-antiquity certificates to articles determined as such and in case of theft, the committee posts 'look out' notices to aid in recovering objects.

Other provisions for international repatriation and aid exist in the form of the 1970 UNESCO Convention on the Means of Prohibiting and Preventing the Illicit Import, Export and Transfer of Ownership of Cultural Property.[11] India ratified this convention in 1977 (United States Information Agency, n.d.) and is also a signatory to the UNESCO Conventions on the Protection of Cultural Property in the Event of Armed Conflict (1955) and Protection of the World Cultural and Natural Heritage (1972). These agreements outside of India, however, do not clarify or offer much help within the country to those who are concerned with the looting of artefacts. However, the repatriation policy mentioned by ASI (being done through Indian Missions in different countries) is mostly unknown.

State initiatives

All Indian states have their own departments of archaeology, with legislation largely patterned on the Ancient Monuments and Archaeological Sites and Remains Act of 1958. The states' constitutional obligations require them to care for monuments other than those nationally listed (Thapar 1984, 66). (Their obligation, however, towards undocumented monuments or antiquity in private hands remains unclear). In some instances, objects cannot be taken into safe custody, especially those associated with religious sites as their removal might compromise their sacredness, akin to 'pretty much the same as being stolen' (Burke 2015, 3).

At the state level, the only authoritative force on relic restitution or theft investigation is the specially appointed Recovery of Idol Wing within the Crime Investigation Department (CID) in Tamil Nadu.[12] This 'Idol Wing' has recovered almost 800 above-ground sculptures in ten years that were taken from remote temples in the state (Burke 2015, 2). Due to the magnitude of the looting problem, however, the Idol Wing is woefully understaffed and has not been able to respond to all complaints of thefts.[13] This effort is made less effective by corruption; recently a CID officer was found guilty of involvement in the illegal export of antiquities (Staff Reporter 2017, 1).

Private investigators

The limited effectiveness of governmental efforts has led to the emergence of groups that use electronic media including blogs and websites to trace missing art treasures.[14] Investigative groups like the India Pride Project (IPP), started by two Singapore-based activists, have aided government agencies in tracking and recovery of objects. Individuals like Vijay Kumar (*Poetry in Stone*, blog https://poetryinstone.in/en/), Anuraag Saxena (co-founder of IPP, https://www.ipp.org.in/) and Dr. Kirit Mankodi (retired archaeologist, *Plundered Past*, blog http://www.plunderedpast.in/fresh-updates.html),

Figure 7.1 Example of a repatriated idol with efforts of Vijay Kumar and the IPP: Ganesha (Bronze). Chola Dynasty, 10–11th CE. Source: The Toledo Museum of Art.

have contributed toward generating unofficial repositories of artefacts aiding to track stolen objects (Figure 7.1), and identifying unscrupulous art dealers. Pursuing undocumented antiquities and the smuggling networks, these groups follow theft reports and scrounge through art catalogues for objects with sketchy provenances. Vijay Kumar states, 'museums continue to shed millions [in lost revenue] due to poor curatorial judgments and faulty acquisition policies, which if, they had properly pursued leads, would have created a good precedent for catching and prosecuting the culprits involved' (Kumar 2017). IPP states, in the history of looting, 'plunder was part of a cultural-conquest, where the winner would want to remove cultural markers', but today it has become a well-oiled global network, a very intricate commercial endeavour (India Pride Project 2017).

Amidst the loot market: the network

The illegal antiquities trade operates much like any economic system based on supply and demand, and continues to thrive on buyers' demand.[15] Recent arrests show that by the time a particular crime gets reported, artefacts have already passed through many hands. The circuitous path involves those on the lowest rung – the daily wage worker who steals for quick money – all the way up to the art gallery owners who get accolades for 'prestigious' finds.[16]

The Kapoor case

In 2011, the arrest of Subhash Kapoor in Germany, a well-known New York-based art dealer, shed light on the inner workings of antiquities' illicit trade. The smuggling network clarifies the journey of the *Suthamalli* idol from Tamil Nadu to New York and the workings of Kapoor's enterprise (Figure 7.2). Kapoor is estimated to have smuggled at least 2,622 items, the majority from India and other Asian countries (Felch 2016). Arrested after an Interpol alert, Kapoor is currently in jail in India on charges of a mere 28 counts of theft, though it has already been established that the art dealer had been running a billion-dollar enterprise that began in the 1980s and is spread across the globe (Boland 2016, 1).

Kapoor's smuggling network relied on thousands of small or remote historic sites scattered all over the country that remain vulnerable to pillage (Figure 7.3). In the absence of proper artefact documentation, obtaining an antiquity clearance is the first step. This is comparatively easy by getting a 'specialist' to sign off on the object as not having any antiquarian value. The second step is to obtain a discretionary certificate from the registered office of the state ASI department. This declares the object 'free' to be exported.

1. Art dealers gets a call/ market demand for particular piece- scouts for unused or unguarded temples or archaeological sites if not in his inventory

2. The objects are usually stolen in the night after the thieves secure the surroundings so that no one can interfere in the theft

3. After being looted these get sent to safe houses of art dealers, far away from site and shop, with many hacked mercilessly while others intact

4. Depending on the site and piece, it either gets ready for transport or spends its time with a local craftsman who makes replicas of the same and floods the market with duplicates

5. In case of export, antiques are disguised amidst modern handicrafts and shipped with fake cargo forms and fake ASI certificates in case of checks

6. Following a circuitous route before they reach their final destination, the antiquities undergo little to no checks on most ports.

Figure 7.2 The loot network, from site to shelf. Source: Author.

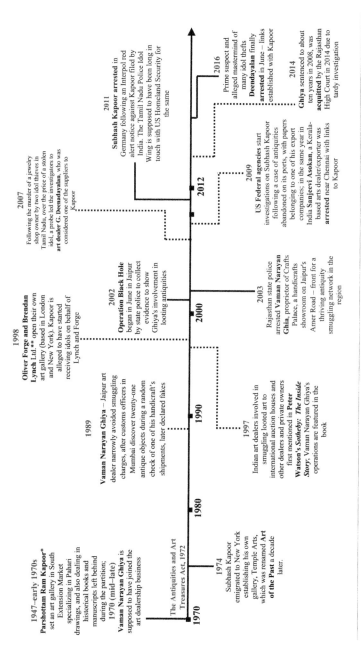

Figure 7.3 Art dealers operating from India and beyond shared a common link of an illicit market. The timeline highlights how they ran in tandem, their years of operation while avoiding arrests, yet eventually getting caught. Source: Author.

Export/import local companies working for international smugglers such as Kapoor, generally send out shipments intermixing antiques with newly crafted traditional wares as part of different cargo and routes to reduce the risk of interception (Mashberg and Bearak 2015, 7). The 2007 case that first put Kapoor under suspicion had an export clearance from the Handicrafts Department of the Union Textiles Ministry labelling the shipments as 'objects of handicrafts' when it fact it contained the Saint Manikkavachakar statue from a temple in Sripuranthan, Chennai, commingled with other idols and modern handicrafts (Mashberg and Bearak 2015, 1).

The case also highlighted the lack of information about the acquirers or collectors who are integral to the smuggling networks. The collectors (many times well-known institutions) will often accept small objects from a lesser known dealer via donations at first, slowly developing a client-business relation based on the artefacts exchanged that eventually creates a new network.[17] Kapoor's arrest also highlights the need for an inventory of artefacts, so that researchers do not need to depend on the old *Indian Archaeology Review* magazines and the French Institute of Pondicherry's photo archive.[18] Museum inventories[19] and Interpol's extensive database created by its art crime wing, based in France, were principal tools used to build a criminal case against Kapoor by the US Homeland Security Department's New York Office of Immigration and Customs Enforcement. Only with the arrest of Govindaraj Deenadayalan, another well-established art dealer (and a Kapoor contact) in July 2016 in Alwarpet, Chennai, did the Idol Wing of Tamil Nadu slowly start making progress, building a case against him, and yet many people in their domestic network remain untraceable.

The Ghiya case

The case of the Jaipur-based art dealer Vaman Narayan Ghiya illustrates how stolen antiquities move from sites to auction houses, traversing a complex network of middlemen unknown to each other. Ghiya would bring Polaroid prints of the object in situ (or after removal from the site) on his trips abroad to entice buyers and would accompany experts from galleries to visit unprotected archaeological sites in India to select artefacts (Keefe 2007, 7). Taking advantage of free trade zones and lax inspections in Switzerland, Ghiya shipped many items via circuitous routes through various shell corporations.[20] These corporations would establish false provenances for the items by buying and selling them multiple times to establish paper trails for each, before they were sold to a customer. Thus, despite different antiquities sometimes having the same mailing addresses,[21] and a small timeframe between each transaction and limited time on foreign soil, the items would attract buyers. The buyers would go on to purchase the items through dummy accounts and paperless money remittance systems (Keefe 2007, 11).

Ghiya was acquitted in 2014 despite a ten-year sentence in 2008 on multiple counts of possession of stolen property and trafficking in looted antiquities. His case exhibits the imbalance between police enforcement and the judicial process. Despite documents and evidence collected in domestic raids by the Jaipur police, jurisdiction issues, absence of other material evidence, and the need to associate an officer from the ASI while searching for antiques,[22] became major loopholes in the case.[23] The local investigating team ignored the fact that Ghiya had already been identified as a key player from India in the international smuggling circuit (Watson 1997, 253). Inter-governmental support, expert advice, and references to prior evidence were also ignored. A 2013 report by the Comptroller and Auditor General (CAG) of India highlights this irregularity, stating that 'the ASI has never participated or collected information on Indian antiquities put on sale at Sotheby's and Christie's as there was no clear provision in the Antiquities Act, 1972 for doing so' (Lal 2015, 2).

The Kapoor and Ghiya cases illustrate that antiques sold on the black market are part of a larger network, where the risk of arrests and fines far outweigh the huge profit margins from successful sales. Enough demand keeps looters busy; local gangs sometimes go to extreme lengths to obtain artefacts; for instance, at the Lal Mandir in Kasba Ramnagar area of Khandauli, Uttar Pradesh, the temple priest was murdered by thieves who escaped with idols made of *ashtadhatu* (also called octo-alloy) in August 2016 (Chauhan 2016).

Scholar and art curator, Pratapaditya Pal, notes that the black market for antiquities remains robust due in part to the Indian laws, that make little distinction between a masterpiece and a generic 'antique', and the government is also lacking in resources to control this market (Keefe 2007, 13). These include lack of measures for constant vigilance (Figure 7.4), task forces behind rigorous database formation, and experts or technical personnel to identify, track, and repatriate items. Mitigation strategies to address these issues should inform good practice for antiquity conservation.

Lessons in mitigation from around the globe

Many nations around the world have invested in resources to prevent illegal excavations, restrict ownership of antiquities, regulate export, and put in place criminal laws to protect archaeological sites (Fincham 2008, 349–351, Mackenzie 2002, 2). In contrast, India struggles to contain theft at both archaeological and cultural sites with ineffectual enforcement.[24] Best practices from other countries, however, can help. In England and Wales, the emphasis is on public awareness and education regarding the historic (not monetary) worth of antiquities, which is achieved through a voluntary system, the Portable Antiquity Scheme (PAS; supported by their 1996 Treasure Act). This system encourages individuals to report finds from undiscovered or unlisted areas. The Act 'requires finders of certain objects to report any find to a designated coroner within 14 days', and the finder is entitled to a reward[25] based on the market price of the object if deemed a treasure of

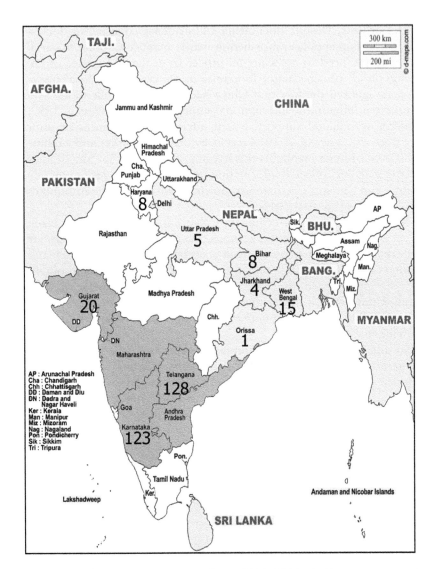

Figure 7.4 The AAT Act arrests from 2012–2014, indicate a pattern in the smuggling racket seen mostly active in the coastal regions of Gujarat and Maharashtra (as well as in the eastern region along Odisha) due to the seaport link. Base map courtesy D-maps, http://www.d-maps.co m/carte.php?num_car=4184&lang=en. Data adapted from TNN. 2016. "Andhra Pradesh is India's hotspot for antiquities theft". *The Times of India*. August 26. http://timesofindia.indiatimes.com/india/Andhra-Pr adesh-is-Indias-hotspot-for-antiquities-theft/articleshow/53868898.cms

the nation. If not reported within the stipulated time, finders are subject to three-month imprisonment, or a fine, or both (Fincham 2008, 352–356). The success of this act indicates an involved public and a slowing demand for such objects.

A similar act might have resonance in India given the ASI's limited resources and the vast distribution of archaeological sites. It may be argued that professional looters could take undue advantage of the act or unprofessional digs could lead to loss of context and probable destruction of the discovery, but a strategic approach could produce results if probable finds were estimated beforehand, keeping in mind the historical site context and the type of object found in the region. Archaeology professionals too could benefit from public awareness, and acts like these would guarantee self-employment for many via activities like rescue archaeology or heritage resource management, encouraging more people to join in saving their heritage (Ota 2010, 94).

Given the finite human resources that are devoted to archaeology, another approach from Peru that utilizes technology to map and monitor in difficult geographical terrain[26] might have potential for India. Since 2008 Peru has been using high resolution satellite data to observe and monitor archaeological sites for looting in their estimated 100,000 plus archaeological areas.[27] Automatic satellite data processing procedures were developed that allow investigators to monitor sites for safety (Lasaponara 2013, 1–14). This use of satellite imagery can be especially effective to monitor remote sites, especially in desert or forested regions in India. Other initiatives like the 'Heritage for Kids' programs undertaken by Heritage Watch, a non-profit based in Australia and the United States, and working in Cambodia, provides a unique approach to generating cultural heritage appreciation and discourage pillage of archaeological sites at a young age. The school curriculum includes a children's book in the local dialect with stories featuring local superstitions to warn against site looting[28] (Thapar 2017). Some models in India are being developed, like the Documentation Resource Centres (under the NMMA) handled in each state by a local college or institute to record antiquities (NMMA 2016, 1), yet many more efforts are needed.[29]

In addition to these models, there is also merit in understanding the role of law enforcement in safeguarding historic and archaeological sites. The Egypt Tourism and Antiquity Police, for example, was created to protect important tourist and archaeological sites in the country (Teijgeler 2013, 230–251). Similarly, the international-level Italian Task Force (ITF) was formulated in February 2016 to protect heritage in conflict zones worldwide. The strategies employed by these task forces could mitigate pillage and vandalism at historic sites in such zones (UNESCO 2016).

The need for a strong institutional network in India with links to the International Council of Museums (ICOM) could benefit management, identification, and other antiquity practices. The ICOM's International Observatory for Illicit Traffic in Cultural Goods, serves as an information databank and network, and also provides tools to deal with the illegal trade. Its Red List includes information on endangered objects or works of art from different countries based on the likelihood of those being traded with the aid of this network (ICOM n.d.).

Extensive inventories along with proper categorization of antiquities could further benefit India in creating a regulated and more open market domestically, like the Japanese antiquity markets, where state ownership of nationally significant antiquities helps identify what stays within, and what can be sold on the open market. Given Japan's regulated market approach, the market evaluations have all come from well-established antique dealers (Conrad 2015, 233-51). Japan's case might offer insights for India. Here it is important to recognize that getting people to declare ownership of what might have been looted items would be a difficult task in India. In this regard, amnesty programs like the 1993 Historic Shipwrecks Amnesty in Australia might provide some leads. That initiative supplemented the database of wreck objects as well as improved knowledge of water archaeology (Rodrigues 2009, 99–102). Moreover, with growing concerns of the antiquity trade financing other illegal activities around the world, trade restrictions like those being devised for the European Union in July 2017, can make it difficult for antiquities from source nation to generate fake paper trails while port-hopping (European Commission 2017, n.d.). Thus domestic amnesties or an open market for better appreciation of antiquities within India might limit the spread of black markets.

With criminal networks often outpacing the efforts of law enforcement authorities, it is imperative for India to be more vigilant and proactive with safeguarding its historic and archaeological sites. A professional network of law enforcement, lawyers and archaeologists in every state can assist with this task, while staying abreast with the market and legislations useful for the country and also work on repatriation of found objects in other parts of the world. There has been sparse coverage in popular media about these thefts, but not enough to raise awareness or develop appreciation. Such losses result in impacting collective understanding of the past and ought to be curtailed by collective and sincere efforts on the part of professionals from a variety of disciplines and members of the community.

Notes

1 The word loot has been derived from the Sanskrit word '*luṇṭhati*'. The word came into common parlance in the 18th century when the East India Company was pillaging from archaeological sites in the north Indian provinces.
2 According to the Oxford dictionary, the word artefact is defined as an object crafted by humans that is of cultural or historical interest.
3 It was noted that ASI updated its Antiquity section in between May- July 2017, to incorporate information on the Data Bank as a subsidiary section along with CAC.

4 According to the ASI website (www.asi.nic.in, accessed July 5, 2017) this was to be achieved within the stipulated time period of 2007-2012; the NMMA website too was updated in June 2017 to accommodate future work.

5 The ASI records a mere 23 accounts of thefts from its protected sites across the country for the 2010-15 period (http://www.asi.nic.in/tenders/D.G.%20fresh2. pdf).

6 As of 2020 there has been little progress in the cases of the identified smugglers, however repatriation has increased. Most recently a 12th century Buddha stolen from Nalanda, a World Heritage Site, has been retrieved from a trade fair in 2018 (along with 14 other stolen statues) and has been returned to India. (http s://www.hindustantimes.com/india-news/india-retrieves-12th-century-brahma-b rahmani-idol-from-london-others-to-follow-soon/story-AQedopXKmJc75JsC9 aThXP.html) Legislative changes are being discussed, but none have been implemented or are available on official websites.

7 First highlighted by Dr. Mankodi on his blogpost; and spotted in an advertisement of a dealer and matched to the posting by Vijay Kumar (Vijay Kumar 2017).

8 The term antiquity as understood from the Antiquities and Art Treasures Act of 1972 can be a work of art or craftsmanship, an article/object/thing detached from a building or cave, of historical importance, illustrative of science, crafts, literature, religion, customs, morals or politics with over a hundred years of existence. It also includes manuscripts with 75 years of existence.

9 The AAT Act forms available on the ASI website (http://asi.nic.in/asi_legisla-tions.asp) state that owner details need to be provided while registering an antiquity, but it is unclear as to whether the forms are only for obtaining license for ownership and selling or extend to documenting private collectors as well.

10 This amount can be appealed later if unacceptable according to the ASI website (http://asi.nic.in/asi_legislations.asp).

11 The convention in a broad sense, requires its state parties to have a structure set up to safeguard their own cultural property, and push for recovering the stolen property while also providing support to other governments when stolen cultural property is found in their country.

12 According to the Tamil Nadu Police Department's Economic Offences Wing website (http://www.tneow.gov.in/IDOL/IW_history.html), the Idol Wing is a police force organized after the increased theft of idols in the 1980s. Others include the lesser known Special Temple anti-theft squad in Kerala and a once constituted special task force under the Central Bureau of Investigation (CBI), seen in the Jaipur case (http://www.keralapolice.org/wings/crime-branch/special -temple-anti-theft-squad, and Keefe 2007, 3).

13 Recently the Idol Wing formed teams to receive complaints of theft from across the state but for a mere two days (https://timesofindia.indiatimes.com/city/tri chy/idol-wing-forms-teams-to-collect-plaints-from-public-about-unreported-t hefts/articleshow/62989850.cms).

14 The Indian National Trust for Art and Cultural Heritage (INTACH) has made provisions for people to report stolen objects or heritage under threat via email, although more information has not been available on the same while writing this chapter (http://intach.org/heritage-risk.php).

15 The artefact is judged primarily on its source of origin, the country/nation. Nations are identified in this system as either a source nation from where objects come or a market nation, where the object is in demand (Fincham 2008, 351). India is identified as a source nation due to its rich historical objects (Mackenzie 2002,1)

16 Unlike a study of the Israel loot market (Kersel 2007, 81-97) there is no detailed data on the Indian market that quantifies the incentives from the looting process (Rodrigues 2009, 92-109) for both looter and local market or source dealer and what other forces drive the looting of archaeological sites or historical structures.

17 Museums and other collectors involved in the Kapoor case have been listed in the *Chasing Aphrodite* (blog). (https://chasingaphrodite.com/tag/subhash-kapoor/).

18 Institute Français de Pondichéry is a 1955 established institution that studies Indian civilization and culture and has a 100,000 plus photo archive devoted to ancient South Indian art (http://www.ifpindia.org/, accessed July 2017).

19 Referring to museums, in which the unprovenanced antique is supposed to be present.

20 A corporation without active business operations or significant assets sometimes used illegitimately, to disguise business ownership from law enforcement or the public. Identified ones for the Ghiya case include Cape Lion Logging, Megavena and Artistic Imports Corporation (http://traffickingculture.org/encyclopedia/case-studies/vaman-ghiya/).

21 As the items would be sold separately, chances of suspicion in provenance checks would not arise as they would hardly ship to the same buyer. In case they did, the similar address would only act as a holding address and not show up on provenance.

22 As required by Section 23 of the AAT Act.

23 The evidence included objects whose antiquary nature was not established nor was the international export thoroughly investigated hence making it circumstantial evidence (http://traffickingculture.org/encyclopedia/case-studies/vaman-ghiya/).

24 It should be noted that the source nation does not always succeed in international repatriations given, its private law is not applicable to the market nation hence both need to be signatory to appropriate international laws.

25 UNESCO noted that 98 percent of the final market price of an object was pocketed by middlemen or as Fincham (2008, 352–356) points out the profits go to the auction houses and dealers. As the final sale value increases a 100-fold from its illegal excavation, adopting rewarding schemes would help prevent it and increase awareness of the law and importance of culture (UNESCO 2011, 4).

26 Conducted by Peru's Ministry of Culture with aid of the Italian Foreign Affairs Ministry.

27 Satellite imagery offers spatial resolution less than 1 meter (considered high resolution data), has a global coverage and allows to re-visit an area multiple times-helping in mapping activity and changes in remote locations.

28 Integrated mainly in the Angkor region, the author heard first-hand accounts of villagers now coming forward to report finds or looting activities in the area via their children.

29 These are merely for recording/registering antiquities under the NMMA mission; others like the Tamil Nadu Hindu Religious and Charitable Endowments (HR&CE) department are commendable efforts but are primarily focused on only bronze and stone idols. The author has not come across efforts where sensitization towards the value of local heritage is ongoing.

References

Archaeological Survey of India. 2017. *Archaeological Survey of India*. Accessed July 5, 2017. http://asi.nic.in/index.asp

Boland, Michaela. 2016. "NGA not dancing for joy at $11.2m Shiva legal win in New York." *The Australian*. September 28. http://www.theaustralian.com.au/arts/visual-arts/nga-not-dancing-for-joy-at-112m-shiva-legal-win-in-new-york/news-story/8b95468bc354ab3be28ea93fd650243b

Brodie, Neil. 2015. Vaman Ghiya. Accessed December 5, 2016. http://traffickingculture.org/encyclopedia/case-studies/vaman-ghiya/

Burke, Jason. 2015. "India struggles to halt multimillion dollar trade in stolen artwork." *The Guardian*. August 3. http://www.theguardian.com/world/2015/aug/03/india-struggles-to-haltmultimillion-dollar-trade-stolen-artworks-temples

Chauhan, Arvind. 2016. "Temple priest found dead, 'ashtadhatu' idols stolen." The Times of India (TNN). August 10. http://timesofindia.indiatimes.com/city/agra/-Temple-priest-found-dead-ashtadhatu-idols-stolen/articleshow/53640586.cms

Conrad, Harald. 2016. "Managing (Un) certainty in the Japanese antique art trade – how economic and social factors shape a market." *Japan Forum* 28 (2): 233–254. doi:10.1080/09555803.2015.1099557.

European Commission. 2017. "Import of Cultural Goods into the EU". European Union. Accessed July 28, 2017. https://ec.europa.eu/taxation_customs/business/customs-controls/cultural-goods_en

Fincham, Derek. 2008. "A coordinated legal and policy approach to undiscovered antiquities: Adapting the cultural heritage policy of England and Wales to other nations of origin." *International Journal of Cultural Property* 15 (3): 347–370. doi:10.1017/S094073910808020X.

Felch, Jason. "Subhash Kapoor tag archives." *Chasing Aphrodite (blog)*. Accessed May 7, 2017. https://chasingaphrodite.com/tag/subhash-kapoor/

Gupta, Gargi. 2016. "India's stolen antiquities: An industrial scale loot." *DNA*. June 19. Mumbai. http://www.dnaindia.com/lifestyle/report-an-industrial-scale-loot-2225242

India Pride Project. E-mail message to author, January 2017.

Indian National Trust for Art and Cultural Heritage. 2012. "Heritage @ risk." *Intach.org*. Accessed March 7, 2018. http://www.intach.org/heritage-risk.php

Institute Français de Pondichéry. 2014. "Photo archive." *Ifpindia.org*. Accessed July 5, 2017. http://www.ifpindia.org/

Keefe, Patrick Radden. 2007. "The idol thief." *New Yorker*. May 7. http://www.newyorker.com/magazine/2007/05/07/the-idol-thief

Kersel, Morag M. 2007. "Transcending borders: Objects on the move." *Archaeologies* 3 (2): 81–98. doi:10.1007/s11759-007-9013-0.

Kumar, Vijay, e-mail message to author, January 2017.

Lal, Neeta. 2015. "Smuggling India's antiquities." *The Diplomat*. September 16. http://thediplomat.com/2015/09/smuggling-indian-antiquities/

Lasaponara, R., Leucci, G., Masini, N. and Persico, R. 2014. "Investigating archaeological looting using satellite images and georadar: The experience in Lambayeque in North Peru." *Journal of Archaeological Science* 42: 216–230. http://www.sciencedirect.com/science/article/pii/S0305440313003798

Mackenzie, Simon. 2002. "Regulating the market in illicit antiquities." *Trends and Issues in Crime and Criminal Justice*. No. 239, September. http://traffickingculture.org/app/uploads/2012/07/regulating-the-market-in-illicit-antiquities.pdf.

Mankodi, Kirit L. 2013. Plundered Past *(blog)*. http://www.plunderedpast.in

Mashberg, Tom and Bearak, Max. 2015. "The ultimate temple raider: Inside an antiquities-smuggling operation." *The New York Times*. July 23. http://www .nytimes.com/2015/07/26/arts/design/the-ultimate-temple-raider-inside-anant iquities-smuggling-operation.html?_r=0 and related article, 2015. "How to smuggle a saint out of India." *The New York Times*. July 23. http://www.nyti mes.com/interactive/2015/07/26/arts/design/kapoor_graphic.html?_r=0

National Mission on Monuments and Antiquities. 2018. "Antiquity search." Accessed July 30, 2018. http://nmma.nic.in/nmma/antiquity_search.do?method =explore_antiquity.

Ota, S.B. 2010. "Archaeological heritage resource management in India." In *Cultural Heritage Management*, edited by Phyllis Messenger and George S. Smith, 1st ed., 82–98. Gainesville: University Press of Florida.

Rodrigues, Jennifer. 2009. "Evidence in the private sphere: Assessing the practicality of amnesties to record lost information." *Archaeologies* 5 (1): 92–109. doi:10.1007/s11759-008-9091-7.

Staff Reporter. 2017. "DSP tainted by idol theft charge suspended: Tamil Nadu." *The Times of India* (TNN). July 1. http://timesofindia.indiatimes.com/city/chenn ai/dsp-tainted-by-idol-theft-charge-suspended-tn/articleshow/59393078.cms

Staff Reporter. 2018. "Idol wing forms teams to collect plaints from public about unreported thefts." *The Times of India* (TNN). February 27. https://timesofindia .indiatimes.com/city/trichy/idol-wing-forms-teams-to-collect-plaints-from-public -about-unreported-thefts/articleshow/62989850.cms

Stein, Deborah. 2011. "To curate in the field: Archaeological privatization and the aesthetic 'legislation' of antiquity in India." *Contemporary South Asia* 19 (1): 25–47. http://cornell.worldcat.org/title/to-curate-in-the-field-archaeologica l-privatization-and-the-aesthetic-legislation-of-antiquity-in-india/oclc/4839416 777&referer=brief_results

Teijgeler, Rene. 2013. "Politics and heritage in Egypt: One and a half years after the lotus revolution." *Archaeologies* 9 (1): doi:10.1007/s11759-013-9231-6.

The Indian High Commission of India. 2016. "India digest." 14 (13), September. London.

Thapar, B.K. 1984. "India." In *Approaches to the Archaeological Heritage*, edited by Henry Kleere, 63–72. Great Britain: Cambridge University Press.

Thapar, B.K. 2017. "Projects." *Heritage Watch International*. Accessed July 5 2017. https://www.heritagewatchinternational.org/projects

Thapar, B.K. 2010–17. "What we do." *ICOM Museum*. Accessed July 15, 2017. http://icom.museum/programmes/fighting-illicit-traffic/red-list/.

Thapar, B.K. 2017. "Cultural goods." *Taxation and Customs Union*. Last modified August 4, 2017. https://ec.europa.eu/taxation_customs/business/customs-contr ols/cultural-goods_en

United States Information Agency. 1998. "1970 UNESCO convention alphabetical list of state parties." International Cultural Property Protection. Accessed July 5, 2017. http://dosfan.lib.uic.edu/usia/E-USIA/education/culprop/unesco03.html

UNESCO. "Illicit Trafficking of Cultural Property". Accessed July 5, 2016. http: //www.unesco.org/new/en/culture/themes/illicit-trafficking-of-cultural-property/ 1970-convention/

UNESCO. 2011. "The fight against the Illicit trafficking of cultural objects, the 1970 convention: past and future." Accessed July 31, 2017. http://unesdoc.unesc o.org/images/0019/001916/191606E.pdf

UNESCO. 2016. "Italy creates a UNESCO emergency task force for culture." Accessed August 1, 2017. http://whc.unesco.org/en/news/1436/

Watson, Peter. 1997. *Sotheby's: Inside Story*. London: Bloomsbury.

8 India's modern heritage
Conservation challenges and opportunities

Priya Jain

Introduction

For the average Indian living either in the teeming urban cities or the rural hinterland of the country, the stripped-down building aesthetic, most commonly associated with the international modern movement in architecture, is hard to escape. Transcending building type, geographic location, and spanning decades, the ubiquity of these modernist buildings demystifies their oft-misunderstood advent with Chandigarh, post-independence nation-building, and western influence. Recent scholarship has charted the rather long arc of 'modernity' in colonial and post-colonial India (Scriver and Srivastava 2015, Scriver and Prakash 2007, Rajagopalan and Desai 2012). It has reframed it in a temporal and social, rather than a stylistic framework. Yet, despite an extended historical understanding, twentieth century–built work in the country is *still* very new (Thakur 2017, 132–134).

With centuries-old heritage that is justifiably threatened, buildings that are barely fifty years old are yet to come of age. Their pervasiveness makes them commonplace; their appeal is judged only visually and they are often appreciated solely by the professional elite. However, these buildings are emblematic of so much more – of massive upheavals in the political, social, cultural, and technological futures of the country. Built over a relatively short span, this 'recent past' is important for much the same reason as most heritage is – to enable a deeper understanding of our place in the present and future. Despite the illegitimacy of age, there are buildings and ensembles from this cohort that merit preservation. Their identification, assessment, and proactive management needs to happen *because* and not in spite of their immaturity. This paper deals with how the modern heritage movement has evolved in India. It notes key milestones and pioneering recent projects at Chandigarh and Ahmedabad, ending with a discussion about future trends in policy and practice in the Indian context.

From 'modern' to 'heritage'

Europe presumably gave birth to modern architecture in a socially progressive atmosphere after World War I, when emboldened by the promise of technology and industrial production, a more egalitarian, functional mode of building was sought and devised. Post–World War II, modernism once again provided a fast, cost-effective way to meet the needs of large populations – social and aesthetic agendas merged to create an omnipresent mainstream style (Betsky 2016, Frampton 2007). America fully embraced modernism after World War II, at a time of phenomenal growth and construction. The colonized nations in Asia, Africa, and Australia had their own indigenous brushes with modernity. Recent scholarship has challenged the oft-repeated narrative of imported western modernity to highlight instead the unique, home-grown 'alternate modernities' that many of these nations actually experienced (Avermaete, Karakayali and von Osten 2010, Lu 2010). Once independent, these fledgling countries embraced modern architecture for both its economic and industrial expediencies but also for its cutting-edge homogenizing aesthetic, well-suited to 'nation-building' (Ashraf and Belluardo 1998, Lang, Desai, and Desai 1997). Initial fascination with modernism, however, gave way to disenchantment; in the late 1970s and '80s, as the western world turned to post-modernism, the once-colonized nations looked back to their pre-colonial pasts to devise regional variants of modernism (Kagal 1986, Scriver and Srivastava 2015).

This was also the time when heritage conservation 'arrived' on the world scene. Becoming a broader populist movement, better integrated with legislative and planning frameworks in the 1960s and '70s, it rose in part, as a reaction to modernist urban renewal projects in the west. The roots of preservation thus ensured a somewhat 'anti-modern' bias to the field (Prudon 2008, 10). It was not until the late 1980s that aging European modern buildings of the 1920s and '30s started to be seen as 'heritage'. Even though DOCOMOMO (Documentation and Conservation of Buildings, Sites, and Neighbourhoods of the Modern Movement) was founded in 1989, it remained a fringe movement for many years, largely concerned with iconic pre-WWII buildings in Europe (Cunningham 1998). The United States soon followed suit with its first 'Preserving the Recent Past' conference in Chicago in 1995 (Slaton and Shiffer 1995). As in Europe, American interest in modern architecture started to evolve slowly from iconic architects and buildings, to everyday vernacular structures and more broad-based research in the conservation of modern building materials and technologies (Jester 1995). As the movement gained steam in the west, some key independent initiatives in India and Asia fanned fledgling interest in twentieth-century built work. In 1999, marking the fiftieth anniversary of Chandigarh, an international conference was organized to celebrate both the legacy and vision of Le Corbusier, and to critically analyse the quintessentially modern city in a socio-economic urban context (Takhar 2002). The discussions were not so much about restoring or protecting Chandigarh or its individual buildings in a strictly conservation sense, but rather evaluating how the modernist design had fared in a developing nation. The discussion also was not about modern Indian heritage in general, but about Chandigarh and Le Corbusier in particular. In Guangzhou, China, a year later, a diverse group of scholars and architects gathered to create a professional group, the 'mAAN- modern Asian Architecture Network' – emphasizing the lowercase 'm' in recognition of diverse perspectives on the issue of modernity and modernism in Asia (Widodo 2010, 79).

A year later, a more concerted international effort, the Program on Modern Heritage, was launched by the UNESCO World Heritage Centre, ICOMOS, and DOCOMOMO for the 'identification, documentation and promotion of the built heritage of the nineteenth and twentieth centuries' (Oers and Haraguchi 2003, 8). ICOMOS also created the International Scientific Committee on Twentieth-Century Heritage (ISC20C) which organized its second regional meeting in Chandigarh in 2003 on modernism in the Asia-Pacific region (Oers 2003). The key issues that emerged from these early efforts were: first, that stylistically modern Asian heritage was of a more 'hybrid' nature; second, identification and protection had to move beyond iconic buildings to vernacular examples and possibly include entire cities (like Chandigarh, Bandung, etc.); third, contextual studies were needed to highlight social, economic, and cultural processes that gave rise to the architecture; and finally, the state of maintenance and adaptation of modern buildings in the developing world raised important questions about authenticity and integrity. It was decided that a 'Conservation Management Plan' for Chandigarh would be developed as a pilot follow-up to the meeting. However, despite successive meetings of UNESCO, ICOMOS, and mAAN, concrete steps towards protection of modern Indian heritage remained scant, particularly outside of Chandigarh.

Not old enough, not special enough

It is important to take a moment to summarize major conservation laws and initiatives in India and discuss whether and how they do (or do not) address modern heritage specifically. Popularly known as the AMASR Act, the Ancient Monuments, Archaeological Sites and Remains Act that has been in place in its current form since the 1950s (and grew out of a colonial era 1904 legislation), forms the legislative basis of architectural conservation in India (Stubbs, Thomson, and Menon 2017, 368). This act, last modified in 2010, retains the colonial-era stipulation that a building be at least one-hundred years old to qualify for protection under its purview. In addition to the national act, many of the states have their own version of the Monuments Act; however, none of these specifically address modern-era buildings. In the absence of relevant national/state legislative tools, some cities such as Mumbai, maintain their own lists of significant modern buildings (Art Deco Mumbai, n.d.). In other cities, like Delhi, local chapters of non-profits such as the Indian National Trust for Art and Cultural Heritage (INTACH), provide advisory lists that serve as the basis of potential legislated protection. These have been prompted in recent years due to the endangered status and demolition of various notable modern buildings – most famously the Hall of Nations in New Delhi in 2017.

Demolition of the Hall of Nations, New Delhi

Designed by noted Indian architect Raj Rewal and built in 1972, the Hall of Nations was considered the world's first and largest space frame structure in reinforced concrete (Langar 2017). Despite its many firsts, association with important architects/engineers, not to mention an eye-catching form, prominent location, and popular appeal, it failed to be recognized as 'heritage'. The flurry of activity in its last days, recounted below, highlights the inadequate protection for modern heritage, but most importantly reveals the philosophical underpinnings of how a building's 'age' weighs heavily in these decisions.

Being the national capital and a union territory, Delhi has a slightly different (more elaborate) heritage designation mechanism than most other Indian cities. Apart from the monuments under ASI control, the Heritage Conservation Committee (HCC) maintains a list of privately owned heritage buildings (Grade I, II, and III). The HCC deliberates over demolitions and alterations to these buildings as well as to non-listed buildings that fall in zones around the ASI-designated monuments (Delhi Development Authority 2016). Then there is an advisory organization, the Delhi Urban Art Commission (DUAC) that advises the HCC and other agencies on matters relating to the overall development of the city. The Delhi Chapter of the Indian National Trust for Art and Cultural Heritage (INTACH) also provides an advisory role.

Leading up to the demolition of the Hall of Nations in 2017, INTACH had prepared 'A Tentative List of Post-Independence Buildings to be notified as Modern Heritage Buildings of Delhi' in 2013, comprising of sixty-two buildings/sites (INTACH 2013) that included the Hall of Nations. After urging the HCC to consider them for inclusion in its roster of listed buildings for many years, they were finally taken up in 2016 only to be determined ineligible. The reasons cited by the HCC varied from a lack of a comprehensive survey, to 'open-ended' criteria for determination that would ensure 'almost any and every building could end up as a heritage building' (HCC 2017). More importantly, it attacked INTACH and DUAC's supposedly 'arbitrary' claim that buildings should be at least fifteen years old for them to be considered as heritage. Instead, it went on to advocate its own unsubstantiated time frame of sixty years (positing that a time interval of at least two generations (thirty years each) as one of the essential conditions for a building to be considered heritage (Chalana 2019, 477–486).

Despite protests from a relatively small national and international community, the Hall of Nations was summarily demolished, underscoring that modern Indian buildings have many impediments to their recognition as cultural heritage (Gupta 2017). Moreover, many of these post-independence

buildings were conceived and built as part of large state-sponsored campaigns, rendering them ploys in a political comeback game with new oppo-
sition parties in power, eager to erase that legacy and supplant it with their
own (Bhatia 2017). The demolition of the Hall of Nations also highlights
the dichotomy that exists between obvious icons of Indian modern architecture built by famous foreign architects, and the more indigenous and
far more prolific versions that exist throughout India. The chapter looks at
concerted conservation efforts at two of these 'iconic' sites and ends with a
discussion on how they bear upon the larger issue of conserving unprotected
modern heritage in India.

Gandhi Bhawan, Chandigarh

As noted above, starting in 1999, Chandigarh started receiving constant
attention from the international and national heritage community—books
were written marking its fiftieth anniversary and major conferences were
organized on site. There was concern about the poor state of the buildings and the Capitol Complex in particular, uncontrolled construction,
and the rampant sale of original furniture pieces on the international market (Garreta 2012, Bharne 2011, 99-112). Encouraged by the inscription
of entire modern cities of Brasilia (1987), Tel Aviv (2003), and Le Havre
(2005), the Chandigarh city administration applied for an unsuccessful
UNESCO World Heritage Site designation in 2007. Finally, almost ten years
later, only the Capitol Complex was successful in being listed, in a completely different transnational listing (with seventeen other nations) as an
exceptional example of Le Corbusier's work (Chalana and Sprague 2013).
The failed city bid and successful Capitol Complex designation are indicative of the fact that modern Indian architecture, when looked at as specific
'iconic' examples, is more readily accepted as 'heritage', rather than more
complete ensembles including 'Ordinary Everyday Modern' (Fixler 2017,
1–11) buildings presenting more complex issues of authenticity, integrity,
protection, and management.

This is borne out also by the recent 'Conservation Plan' for Gandhi
Bhawan, a 1962 building by Pierre Jeanneret situated on the Panjab
University campus in Chandigarh (DRONAH 2017). Completed as part of
the Getty Conservation Institute's 'Keeping It Modern Grant' initiative, the
comprehensive two-year planning was intended not only to chart a course
of conservation for the well-known building, but also serve as a prototype for conservation of other modern-era buildings in the nation. Led by
DRONAH, a Gurugram-based Indian conservation firm, the plan finished
in 2017, and was able to draw on the experience of both local/national
experts and international partners and case studies.

While preservation planning has been employed in various ways
throughout the twentieth century, and most systematically in the US after
the 1966 National Historic Preservation Act (NHPA), the more specific use

of 'conservation plans' as a tool to guide interventions at historic properties can be traced to Australia in the 1980s (Kerr 1982). While the term and methodology has significantly evolved in ensuing years, the basic tenets of a 'conservation plan' remain the same – identification of significant historic attributes of a building, site, or resource and framing of specific policies that ensure its ongoing stewardship. In terms of modern buildings, the first and most well-known of such plans was prepared for the Yale Center for British Arts (YCBA), a 1974 Louis Kahn building in New Haven, Connecticut (Inskip and Gee 2011). Prepared by British conservation architects, Peter Inskip and Stephen Gee, it charted the way for numerous other conservation plans, like the one for Gandhi Bhawan, that have been funded by the Getty since 2014 (MacDonald et al. 2018, 62–75). Each conservation plan is tailored to the particular resource and cultural context. Accordingly, the Gandhi Bhawan plan focuses more heavily on training and capacity building of staff and professionals, includes a risk management component, pilot projects, a landscape plan, an interpretation and use plan, as well as 'secondary' plans for more short-term implementable projects. Much like similar projects in the west, there was a lot of emphasis on scientific testing and diagnostics – in this case, of the concrete façade and waterproofing of the surrounding pool. These studies are key to establishing procedures for other modern-era buildings in India. The conservation plan also included digitization of the archival drawings, extensive research on building materials, and a detailed inventory and assessment of all the original furniture. The Conservation Plan for Gandhi Bhawan is both thorough and comprehensive in scope – but falls short on including a clear mechanism for funding the projects outlined, and for updating the plan itself a few years down the line. It also leaves one hoping that the conservation planning could be extended to the larger Panjab University campus, designed by Jugal Kishore Chowdhury and Bhanu Pratap Mathur, significant Indian architects in their own right, in collaboration with Jeanneret. This would also allow for assessing and potentially conserving the remaining campus, that not only provides context to Gandhi Bhawan, but is equally, if not more, important from a socio-cultural perspective.

Indian Institute of Management, Ahmedabad

The campus of the Indian Institute of Management (IIM-A) in Ahmedabad was designed by renowned American architect Louis Kahn and built from 1964–75. An impressive ensemble of exposed brick buildings with numerous courts and verandahs (see Figure 8.1), it is emblematic of the emerging regionalism within Indian modern architecture of the 60s and 70s. Starting in 2014, roughly at the same time as the Gandhi Bhawan Plan, similar conservation planning was initiated at IIM-A with two key differences. First, the project scope covered not just one building but the entire Kahn campus – dormitories, academic buildings, and faculty blocks. Second, the funding

Figure 8.1 Indian Institute of Management-Ahmedabad, built 1964–75, photo 2016. Architect: Louis Kahn. Source: Perspectives – The Photography Club, IIM Ahmedabad / CC BY-SA, https://commons.wikimedia.org/wiki/File:Louis_Kahn_Plaza,_IIM_Ahmedabad.jpg

for the initiative came not from international agencies but from a national public-private partnership (Somaya 2018). This was likely due to the more market-driven focus of the management institute and the fact that the Kahn name and iconic campus architecture has become an integral part of its brand. Moreover, following earthquake damage in 2001, the IIM-A buildings were in a dire condition and needed urgent repairs. Technically, the conservation planning process followed a similar approach as Gandhi Bhawan – there were extensive surveys, archival research, and large-scale mock-ups before selecting the first buildings for renovation. At IIM-A, easier access to funding has ensured that actual conservation work has proceeded more readily – the library building and one of the dormitories were the first to be renovated in 2018 (Somaya 2018, 66–70). This is proposed to be followed by the remaining dormitories, faculty block, and classroom buildings.

The restoration of these buildings at IIM-A has enabled some key outcomes –first, the meticulous matching of brick, archival research, investment in mock-ups, etc., has raised the profile and level of care that maintenance staff, users, and other professionals afford to the buildings. The successful integration of new systems, functional program requirements, and sympathetic interior finishes has demonstrated that heritage conservation and new adaptations can indeed go hand in hand. It is also hoped that this will deter the campus authorities from pursuing new replacement buildings and see the value in retention, adaptive use, and revitalization of existing structures. Much like Gandhi Bhawan, capacity building of the institute staff and local building crew, and comprehensive construction documentation has enabled that restoration of the remaining dormitory buildings can be carried out directly by IIM-A, while the specialized conservation professionals can focus on creating a prototype for the faculty and classroom buildings (Somaya 2018). Some campus-wide guidelines and policies established as part of the conservation plan – such as pointing, brick cleaning and rebuilding, seismic reinforcement, etc., can be directly extrapolated to the remaining buildings. In commenting on the conservation, Brinda Somaya, of SNK Architects, the Bombay-based firm behind the project, noted that by being the first high-profile modern heritage project in India, IIM-A has the potential to raise the profile of similar twentieth-century buildings. Yet, she conceded, that for a country with centuries-old heritage, modern buildings will continue to have to establish their relevance, find funds, and win popular support (Somaya 2018).

Indian modern heritage: legacy, appeal, and relevance

The meticulous and systematic planning at Gandhi Bhawan and IIM-A are promising prototypes. Conservation master planning allows both a proactive diagnostic model as well as a long-term management tool. There are a variety of other Indian sites where a similar model can be deployed and integrated with other planning activities already in place – a case in point would be the five campuses of the Indian Institutes of Technology (IIT). Built between 1951 and 1963, these quintessentially modern campuses are representative of the fledgling nation's investment in scientific education during the post-independence decades. Their design and construction provide a telling narrative on the 'coming of age' of the Indian architectural profession. IIT-Kharagpur, the first to be built in 1951, initially engaged a foreign campus planner. By the mid-1950s, at both IIT-Kharagpur and IIT-Madras, Indian architects who held proprietorship roles at Anglo-Indian firms were handling these commissions (J.K. Gora and A.K. Bose at IIT-Kharagpur and B. Pithavadian at IIT-Madras). Finally, by the 1960s when IIT-Kanpur and IIT-Delhi were planned, young Indian architects with no colonial ties and a strong modernist bent (A.P. Kanvinde at IIT-Kanpur and J.K. Chowdhury at IIT-Delhi, see Figure 8.2), were selected for these massive projects. Thus, the IITs shed a unique light on the mainstreaming of modernism in the Indian milieu. Each of these campuses, now more than fifty years old, is also faced with aging buildings, land-locked sites, and the need to maintain a 'state-of-the art' image. Conservation master planning will enable them to better acknowledge the 'heritage' aspect of their sites and deter decisions that advocate for demolition and insensitive infill construction.

And yet, the IITs present but just a drop in the ocean of significant modern buildings/campuses in India—the work of B.V. Doshi (at the Indian Institute of Management, Bangalore [1962–74] as shown in Figure 8.3, and the Centre for Environment and Planning Technology, Ahmedabad [1972],

Figure 8.2 Indian Institute of Technology-Delhi Main Building, built 1960–69, photo 2018. Architect: Jugal Kishore Chowdhury and Gulzar Singh. Source: Author

among other projects), has recently got renewed attention after his Pritzker Prize award in 2018. Other Indian architects like Charles Correa, Joseph Allen Stein, and Laurie Baker (the latter two being foreigners who spent the majority of their careers in India) are also recognized broadly. Yet, there are many others who are barely known outside Indian architecture circles: Iftikar Kadri, Kuldip Singh, and Leo Pereira to name a few.

Lessons for practitioners

As borne out by the discussion above, the extension of conservation planning to other modern-era institutional and cultural campuses, though largely absent at the moment, seems like a desirable goal – yet, in the larger context, it is still a low-hanging fruit. The larger issue confronting modern heritage in India is the sheer volume and diversity of the typology (Chalana 2019). To that end, one key priority ought to be the creation of a national thematic context for modernism – how it evolved, manifested itself, and continues to have relevance in the Indian milieu. The Getty Conservation Institute in partnership with ICOMOS identified the need for an international thematic context for modern architecture in 2011 (MacDonald and Ostergren, 2011). A 'thematic context' identifies key trends and larger values that

Figure 8.3 Indian Institute of Management – Bangalore Academic Block, built 1973, photo 2006. Architect: B.V. Doshi. Source: Sanyam Bahga / CC BY-SA, https://commons.wikimedia.org/wiki/File:IIM-B_016.jpg

characterize a particular cultural resource group, and enables identification of representative examples that portray those themes.

At an international level, some of these themes for modern heritage include urbanization, mass communication, transportation, internationalization, etc. A parallel, national effort is needed to identify themes that would define post-independence built work in India: science and technology education, campus building, industrial infrastructure, government housing, tropical modernism, and so on – trends that draw from but are unique to the special social, cultural, and political forces that shaped architecture in post-independent India. This will allow a broader appreciation of modern buildings, not only stylistically but also because of their association with these larger themes, hopefully extending their popular appeal and paving a path for their eventual legislative protection at local and state levels. It will also allow the shift from 'iconic' to everyday examples (like the nondescript commercial buildings at Nehru Place in Delhi (see Figure 8.4) and Nariman Point in Bombay), while allowing the critical framework to sort through the sheer volume of modern buildings. It will hopefully dispel the perception of modernism as a western import – deriving its sole significance from association with foreign architects, highlighting instead and in addition, the work of lesser-known post-independence Indian architects.

Figure 8.4 Nehru Place – Delhi, c.1980, photo 2010. Delhi Development Authority.
Source: Adam Geitgey / CC BY-SA, https://commons.wikimedia.org/
wiki/File:Nehru-1.jpg

Conclusion

While the modern heritage movement in the west has been dominated by discussions around materiality, finishes, and challenges presented by the experimental, non-durable nature of twentieth-century buildings, the focus in India ought to be wider and not guided by recreating the 'clean', un-weathered image. It needs to acknowledge modernism not simply as a historical movement but one that continues to find contemporary relevance – via adaption, replication, and reinterpretation of its simple geometries and economical modes of construction. To enable this 'recent past' to stand up against centuries-old monuments and more obvious candidates for conservation, its place in the 'story of India' needs to be better studied and disseminated. The recent inscription of the Victorian Gothic and Art Deco (Indo-Deco) ensembles of Mumbai on the World Heritage List signal that an appreciation of the late nineteenth- and early twentieth century-built work in the country is on the horizon. One hopes that it will diffuse beyond the metros to other cities and smaller towns, and most importantly, outpace misguided development and demolition.

Bibliography

Art Deco Mumbai. n.d. "Art Deco Mumbai." http://www.artdecomumbai.com/, Accessed December 1, 2018.

Ashraf, Kazi K. and James Belluardo. 1998. *An Architecture of Independence: The Making of Modern South Asia: Charles Correa, Balkrishna Doshi, Muzharul Islam, Achyut Kanvinde.* New York: Princeton Architectural Press.

Avermaete, Tom, Serhat Karakayali and Marion von Osten, eds. 2010. *Colonial Modern: Aesthetics of the Past—Rebellions of the Future.* London: Black Dog Publishing.

Betsky, Aaron. 2016. *Making it Modern.* New York: Actar Publishers.

Bharne, Vinayak. 2011. "Le Corbusier's Ruin: The Changing Face of Chandigarh's Capitol." *Journal of Architectural Education* 64 (2): 99–112.

Bhatia, Gautam. 2017. "Break in India." *India Today*, May 5. https://www.indiatod ay.in/magazine/up-front/story/20170515-delhi-architecture-monuments-history -infrastructure-986328-2017-05-05

Chalana, Manish. 2019. "The Future of the Recent Past: Challenges Facing Modern Heritage from the Postcolonial Decades in India." In *Routledge Companion to Global Heritage Conservation*, edited by Vinayak Bharne and Trudi Sandmeier, 477–486. London: Routledge.

Chalana, Manish and Tyler S. Sprague. 2013. "Beyond Le Corbusier and the Modernist City: Reframing Chandigarh's 'World Heritage' Legacy." *Planning Perspectives* 28 (2): 199–222.

Cunningham, Allen. 1998. *Modern Movement Heritage.* New York: Routledge.

Delhi Development Authority. 2016. "Unified Building Bye Laws for Delhi." Annexure II: Conservation of Heritage Sites including Heritage Building, Heritage Precincts and Natural Feature Areas. Accessed December 15, 2018. http://www .indiaenvironmentportal.org.in/files/file/UBBL_Delhi%202016.pdf

DRONAH. 2017. Conservation Management Plan for Gandhi Bhawan. Panjab University. https://www.getty.edu/foundation/initiatives/current/keeping_it_m odern/report_library/gandhi_bhawan.html, Accessed January 2, 2019.

Fixler, David N. 2017. "Introduction to the Special Issue." *Journal of Architectural Conservation* 23 (1–2): 1–11.

Frampton, Kenneth. 2007. *Modern Architecture: A Critical History.* London: Thames & Hudson.

Garreta, Ariadna Alvarez. 2012. "Chandigarh Heritage Furniture." *Docomomo Journal* 47 (2): 74–79.

Gupta, Narayani. 2017. "With Custodians Like These." *Indian Express*, February 22. https://indianexpress.com/article/opinion/columns/agencies-charged-with-the -upkeep-and-protection-of-delhis-historic-buildings-find-little-thrill-in-their- job-heritage-4536883/?utm_medium=website&utm_source=archdaily.com

HCC. 2017. "Minutes of the 53rd Meeting of the Heritage Conservation Committee Held on February 2, 2017. No. 2(1)/2004-HCC." https://architexturez.net/doc/ az-cf-182386, Accessed January 15, 2019.

Inskip, Peter and Stephen Gee. 2011. *Louis I. Kahn and the Yale Center for British Art.* New Haven: Yale Center for British Art.

INTACH. 2013. "A Tentative List of Post-Independence Buildings to be Notified as Modern Heritage Buildings of Delhi." https://architexturez.net/doc/az-cf-1667 47, Accessed December 20, 2018.

Jester, Thomas C. 1995. *Twentieth-Century Building Materials.* New York: McGraw-Hill.

Kagal, Carmen, ed. 1986. Vistāra: The Architecture of India, Catalogue of the Exhibition. Bombay: The Festival of India.

Kerr, James Semple. 1982. Conservation Plan: A Guide to the Preparation of Conservation Plans for Places of European Cultural Significance. Sydney: National Trust of Australia.

Lang, Jon T., Madhavi Desai and Miki Desai. 1997. *Architecture and Independence: The Search for Identity India 1880–1980*. Delhi: Oxford University Press.

Langar, Suneet Z. 2017. "The Demolition of Delhi's Hall of Nations Reveals India's Broken Attitude to Architectural Heritage." *ArchDaily*, June 23. https://www.arc hdaily.com/874154/the-demolition-of-delhis-hall-of-nations-reveals-indias-b roken-attitude-to-architectural-heritage/

Lu, Duanfang, ed. 2010. *Third World Modernism: Architecture, Development and Identity*. New York: Routledge.

MacDonald, Susan, Sheridian Burke, S. Lardinois and C. McCoy. 2018. "Recent Efforts in Conserving 20th-Century Heritage: The Getty Conservation Institute's Conserving Modern Architecture Initiative." *Built Heritage* 2: 62–75.

MacDonald, Susan, and Gail Ostergren. *Developing an Historic Thematic Framework to Assess the Significance of Twentieth-Century Cultural Heritage: An Initiative of the ICOMOS International Scientific Committee on Twentieth-Century Heritage*. Los Angeles: The Getty Conservation Institute, 2011. Accessed December 11, 2018.

Oers, Ron van. 2003. "Second Regional Meeting on Modern Heritage: Asia. Chandigarh (India), February 24-27, 2003." *DOCOMOMO Journal* 29 (September): 17–18.

Oers, Ron van and S. Haraguchi, eds. 2003. *Identification and Documentation of Modern Heritage*. Paris: UNESCO World Heritage Centre. https://whc.unesco. org/document/3194/, Accessed November 7, 2018.

Prudon, Theodore H. M. 2008. *Preservation of Modern Architecture*. Hoboken: Wiley.

Rajagopalan, Mrinalini and Madhuri Desai, eds. 2012. *Colonial Frames, Nationalist Histories: Imperial Legacies, Architecture, and Modernity*. Surrey: Ashgate.

Scriver, Peter and Vikramaditya Prakash, eds. 2007. *Colonial Modernities: Building, Dwelling and Architecture in British India and Ceylon*. New York: Routledge.

Scriver, Peter and Amit Srivastava. 2015. *India: Modern Architectures in History*. London: Reaktion Books.

Slaton, Deborah and Rebecca A. Shiffer, eds. 1995. *Preserving the Recent Past*. Washington: Historic Preservation Education Foundation.

Somaya, Brinda. 2018. Restoration of IIM-A, Interview by author, November 16.

Stubbs, John H., Robert G. Thomson and A. G. Krishna Menon. 2017. *Architectural Conservation in Asia*. New York: Routledge.

Takhar, Jaspreet, ed. 2002. *Celebrating Chandigarh: 50 Years of the Idea*. Ahmedabad: Mapin Publishing.

Thakur, Nalini. 2017. "India." In *Time Frames: Conservation Policies for Twentieth-Century Architectural Heritage*, edited by Massimo Visone and Ugo Carughi, 132–134. London: Routledge.

Widodo, Johannes. 2010. "Current State of Modern Asian Architecture Discourse and Networking." *Journal of Architectural Education* 63 (2): 79–81.

9 Heritage conservation and seismic mitigation in small-town India

The case of Chamba, Himachal Pradesh

Manish Chalana and Sakriti Vishwakarma

Introduction

The practice of heritage conservation in India continues to be both urban- and monument-centric. Additionally, the bulk of cultural resources remain unprepared for natural disasters, even in hazard-prone settings. With rapid loss of traditional buildings that generally exhibit better risk preparedness, their modern replacements pose a greater threat to loss of life and existing historic buildings. This chapter focuses on one such town, Chamba, in the Chamba Valley of Himachal Pradesh, which is located in a high seismic risk zone and remains prone to landslides, flash flooding, and forest and urban fires. This chapter examines the range of cultural resources in the town and its environs, including monumental and vernacular cultural resources as well as cultural landscapes. It argues for the value of vernacular building traditions and cultural landscapes in placemaking and response to natural hazards. Chamba is representative of numerous small towns in the Indian Himalayas that need a comprehensive cultural resource management plan that is informed by disaster mitigation strategies to ensure their historic continuity in the 21st century.

Chamba town

Chamba is located at the confluence of the Ravi and Sal Rivers in the inner Himalayas, over 300 km south-east from Shimla. Historically an agrarian town, Chamba today serves as the headquarters of Chamba District with a predominantly Hindu population (~90 percent) of roughly 20,000 (Directorate of Census Operations [H.P.] 2011). Post-Partition of India in 1947, Chamba was part of the Indian state of Punjab which witnessed a major influx of refugees (Negi 1963). Eventually, Chamba would be included in the new state of Himachal Pradesh, much of which was carved out of East Punjab (Negi 1963). In the decades following partition, the state initiated a range of government-led initiatives including hydroelectric projects that ushered in development and population growth in the region (Negi 1963). The Chamba region falls in a high-risk seismic zone that poses risks to life and safety as well as the historic built and natural environments. In recent decades, the town's sporadic growth has brought in urban problems that are testing its carrying capacity and threatening the unprotected cultural resources. If such transformations continue unchecked, particularly in the absence of seismic mitigation strategies, the town is likely to lose most opportunities to maintain its cultural resources and traditional practices in any meaningful way for residents and visitors alike.

Sacred cultural landscapes

Himachal Pradesh (HP) is considered *Dev Bhumi* or the abode of Gods; it is believed to be the home of Shiva, who is part of the Hindu trinity along with Vishnu and Brahma. The entire region is sacred to Hindus, particularly Shaivites (followers of Shiva) and has been shaped through millennia by myths, beliefs, and ongoing traditional practices, which are reflected in the built fabric (including open spaces), as well as in rituals, folklore, fairs, and festivals. The locals engage physically, mentally, and spiritually with such places, which can include a range of natural and cultural landscapes: mountains, rivers, pools, lakes, groves, towns, and villages (Eck 1993; Sinha 2006). Today, social life continues to revolve around a variety of rituals and practices in many villages and small towns of HP. As Sinha argues, sacred landscapes in India are regarded as "places shaped by a way of seeing the divine in nature and physically engaging with it through ritual activities. These natural archetypes "symbolize the axis mundi, a link between the earth and heavens, where divine encounters are most likely to occur" (Sinha 2017).

The Chamba Valley remains a part of the "Western Himalayan Cultural Complex" with its "diverse range of understandings of the sacred [including] belief in a range of denizens of the spirit world" (McKay 2015, 161). It spreads across the mountains between West Nepal and Kashmir and has been defined as "elements of the regional culture that predate, fall outside, or have survived the impact of World-religions" (McKay 2015, 153). Several belief systems associated with renunciation, enchantments, witchcraft, and body possessions all remain common in HP. Temples and shrines help strengthen connections to gods and keep evil at bay; they are typically unassuming structures located within the town and outside in non-urban settings as well (Shashi 1971).

The sacred geography of the region of Chamba is tied to *Kaplas* (Mount Kailash). Shiva is believed to reside at the base of Mount Kailash by Manimahesh Kailash (or lake), 40 miles east of Chamba town. The lake and the mountain are both sacred; a yearly fair and pilgrimage (or *yatra*) in *Bhadon* (or the 31-day long season in August/September) is held at the Laxmi Narayan group of temples (discussed later); the *yatra* commences from there and concludes at the *Kaplas* Kailash in which thousands of pilgrims participate (Bharti 2002; 1989; Shashi 1971). The sacred geography of this region is also linked to the history of Shaivites (also known as *jogies*) who give up worldly trappings for spiritual pursuits, and *naga* worshippers (*nagas* or cobras are associated with both Vishnu and Shiva). Within sacred landscapes are sites that are unique and meaningful to communities with ties to traditional cultural practices. These sites are especially imbued with religious or spiritual meaning and are often marked by built structures (Bhatnagar 2008; Charak 1978; Mehta 2011). In Chamba, such markings take the form of small temples or shrines in various neighbourhoods, and along the town's periphery marked by a string of temples that collectively form a protective ring (or a *parikrama*) to ward off evil spirits.[1]

The setting of Chamba includes terraced fields; those in the lower terrain are used for cultivation of rice, while other crops such as barley, maize, and sugarcane are cultivated on the upper terraces (Institute of Integrated Rural Development 2011). Much of the agriculture in Chamba Valley relies on water drawn from perennial streams and rivulets through channels, pools, canals, or cuts (Deambi 1985; Khandalavala 1989). The historic *kuhl* irrigation system is gravity-fed, similar to other regions in the Himalayas; it is a rather complex system consisting of a hierarchy of main channels, numerous lateral channels, diversion structures, and distribution points. Prior to rural electrification, these *kuhls* also provided hydropower for grain milling and water for domestic purposes (Baker 1997). Both agricultural lands and water systems are shared among groups based on kinship networks, who share their maintenance as well as yields and profits. The *kuhls* are managed under the leadership of a *kohli*, who is appointed to oversee maintenance and repair and facilitate rituals associated with water management. Such collective land management practices are particularly suited for harsh mountainous environments as they are better able to balance the system's carrying capacity and community needs (Baker 1997).

Nature is worshipped in many forms in the Chamba Valley. *Varuna* – the God of Water – has a special significance since water is sacred, scarce, and a common property. *Varuna*'s figure is prominently displayed in the centre of the carved stone on most fountain slabs. This stone is nearly unique to this region; the only other place it is found is in Sisu in Lahaul (Hutchison 1904; Hutchison & Vogel 1933/1982). The fountains ensure access to drinkable water from underground streams in neighbourhoods. It is believed that patrons who invest in communal resources such as the water fountains accrue spiritual merit in their next life. Almost all forms of nature are sacred, but groves and forests associated with a temple rank especially high. In addition to their ecological merits of protecting local (and sometimes rare or endangered) species of flora and fauna, forests also provide firewood and fodder for domestic animals as well as fruits and medicinal herbs (Anthwal et al. 2010).

Chamba town morphology

The traditional settlements in Himachal Pradesh follow a similar urban form. The built fabric is crafted out of locally sourced building materials, primarily stone and timber, using traditional building techniques. Most towns are low-lying and compact where urban form has to navigate difficult terrain. This is achieved by creating terraces using the cut-and-fill techniques common in hilly terrains. The traditional towns typically have a large open space for religious and cultural activities, and a bazaar along which retail and manufacturing is organized. The most prominent landmarks in such towns are temples and/or a palace. The neighbourhoods in these towns are organized as *mohallahs* based on kinship and caste networks. Such settlements evolve organically out of the folds of the landscape, reflecting purpose and expression through traditional building practices and construction techniques (Dave et al. 2012).

The town of Chamba grew around the palace and temple complex first toward the east, and eventually toward the west and south during the British colonial period in the late 19th century, made possible by a vehicular access road and a suspension bridge[2] to the south (Hutchison & Vogel 1933/1982). It is located on a plateau above the junction of the Ravi River and its tributary Sal in the lower valley. The elongated town navigates the hilly terrain in three distinct zones, the lowermost of which is largely uninhabitable due to its steep terrain. The middle zone has a relatively flat terrain (known as *Chowghan*), which is a public space for fairs and festivals. The upper zone with a gentler gradient is where the bulk of the township is located.

Protected historic sites

Shri Laxmi Narayan group of temples

Adjacent to the palace is the royal temple complex – Laxmi Narayan – with six *Nagara* style temples distinguished by their curvilinear spires, three each dedicated to Shiva and Vishnu. The temples are aligned along the east-west axis and are based on the *vastupurusha* mandala concept of temple planning. The *Nagara* stone architecture in the Chamba Valley has influences from Kashmir; the structures are built of layers of stone masonry alternating with beams of deodar wood (Meister 1979). Elaborate and detailed carvings adorn the facades, ceilings, and columns of the temples (Figure 9.1). Extended porches or *mandapas* with pent roofs are later additions (Goetz 1955; Hāṇḍā 2001; Hutchison 1904). In Chamba, temples are a vital part of everyday life; town residents often stop by at these temples to seek blessings on their way to and from work. The Shri Laxmi Narayan group of temples is protected by the Archaeological Survey of India (ASI) as a national monument and managed by the temple trust.

Unprotected vernacular heritage

Akhand Chandi Palace complex

Above the *Chowghan* lies the Akhand Chandi Palace, the royal residence built in the 18th century by Raja Umed Singh (1748–1764 CE). The palace

Figure 9.1 Shri Laxmi Narian group of temples; view from the southern corner of temple complex. Source: Sakriti Vishwakarma

uses a traditional *chatushala* form comprising four structures arranged along a courtyard linked by a hallway on the upper floor. During the colonial period, Zenana Mahal and Darbar Hall were added. The new buildings exhibit both Mughal and colonial influences[3] in their use of baked bricks, cusped arches, *jaali* patterns, and ornamentations. The palace buildings are currently in use as an institution of higher learning, Government College Chamba, and can be accessed from the *Chowghan* via a set of steep and winding stairs. The palace remains the most dominant structure in the town.

Residential mohollahs[4]

Much of the town outside of the *Chowghan*, palace, and temple complex is organized as residential neighbourhoods or *mohollahs* based on kinship and caste groupings. Typically, in a *mohollah* typology the location and names indicate the class and caste hierarchy of residents (Bandarin et al. 2011; Hosagrahar 2001; Hosagrahar 2002). There are 13 *mohallahs* in the historic town of Chamba, which have a distinct social hierarchy based on their location and built form. Those located on higher topography are older and continue to be inhabited by the more affluent and upper-caste residents, and exhibit elaborate vernacular built form. Two *mohallahs* adjacent to the elite *mohallahs* were traditionally home to *dhobis* (laundrymen) and *bhishtis* (road cleaners) who historically served the upper-class residents. Other non-elite *mohallahs* located in the lower terrains are newer and inhabited by

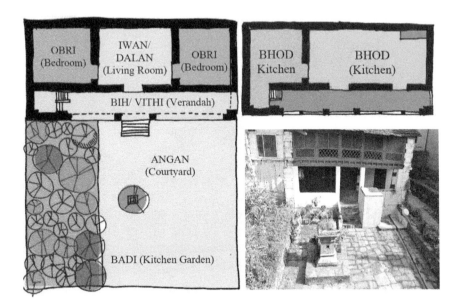

Figure 9.2 Ground-level plan of a *shala* (left); first-level plan (above right). The image (below right) shows a typical *shala* structure in a *thatara tola* construction in Chamba, with a *badi* and *tulsi vedika* in the front courtyard. The four concrete columns in the front veranda, and a small outdoor bathing area on the right of the entrance steps are more recent additions. Source: Sakriti Vishwakarma; sketch generated in September 2014 as part of a studio project at the School of Planning and Architecture, New Delhi

members of the working classes and lower castes, refugees, and more recent arrivals (Town and Country Planning Department H.P.). All *mohallahs* (as well as the *Chowghan*, temple, and palace complex) are networked by a hierarchy of streets, of which only some can support vehicular traffic. The house layout in the *mohallahs* can take multiple forms based on the arrangement of a single structure (or *shala*), which functions as an irregular module for spatial arrangement. Figure 9.2 illustrates a *shala*, which is a double-storey compact structure with living and bedrooms on the lower level and kitchen on the upper level connected by a staircase from the outside. *Ekshala* or a single structure could have a courtyard in the front or on the sides; *dvishala* has two *shalas* arranged at right angles or across from one another facing the courtyard; *trishala* has three *shalas* arranged in C-shape; and *chatushala* would have four *shalas* arranged around a courtyard. The greater the number of *shalas*, the higher the caste (and class) of the resident. Typically, the courtyard was marked by a *tulsi vedika* (or holy basil) plant and included a kitchen garden (*badi*) and some fruit trees. The courtyards continue to be used as outdoor rooms for everyday life practices, rituals, and leisure.

Public spaces

Today the *Chowghan* is the town's largest open space, used for religious, cultural, recreational, and social events throughout the year. It was expanded and levelled in a rectangular shape for use as a parade ground during the colonial period. A colonnaded promenade of shops and prominent administrative structures were built along its edge to define the public space (Town and Country Planning Department H.P.). Vernacular shop-houses line the street connecting the *Chowghan* to the palace. The lower floor of the shop-houses is raised on a stone plinth that can be accessed directly by steps from the street. The upper floors are used as residences and/or workshops of shopkeepers; they have projected wooden balconies resting on timber planks that create a decorative awning (or *chajja*) offering protection from weather. Similar shop-houses are also found across from the Laxmi Narayan Temple complex that once specialized in temple-related wares, but are now more general purpose retail establishments.

Traditional construction techniques

While the vernacular in Chamba is not that dissimilar to the other Himalayan regions, it is unique in the use of locally sourced materials and responses to topography and climate. The construction technique in the Chamba region is recognized as *thatara tola*, where; *thatara* is a local term for timber planks that make the vertical load-bearing columns, and *tolas* is the filling of random stone rubble that enhances seismic preparedness illustrated in Figure 9.3 (Rahul et al. 2013). The walls are built of alternating courses of wooden planks and tightly packed loose stones (historically no mortar was used in the stone course). They are finished with a mix of sand, fibre (straw and thatch) and cattle dung plaster, which insulate the interiors from harsh winters. In some houses both the interior and exterior walls (made of stone and wood) are plastered; and interior walls sometimes display coloured wall paintings depicting scenes from Hindu mythology.

Hybrid vernacular urban form

The British organized Chamba under the Punjab Province after the First War of Independence (aka "Mutiny") in 1857. They saw Chamba as "a relic of the past" and took great interest in its antiquity,[5] but at the same time began modernizing the town by investing in public works improvements including construction of roads, police stations, a court house, and a state hospital. The buildings from the British period are not entirely European in style but are instead an amalgamation of local, Mughal, and English styles. They display arches with decorative patterns similar to Mughal multifold arches and medallions found in Delhi and Agra, and use Doric columns and stained glass windows. In addition, they sometimes exhibit dressed stone masonry,

Figure 9.3 View from the Rang Mahal showing *Chowghan* on low grounds to the
left at the base of the hill; in the foreground are Bangotu and Darobi
mohollahs showing a mix of old and new urban form; on the right high
elevation is Akhand Chandi Palace. Source: Sakriti Vishwakarma

sloping roofs with deep verandas and facades wrapped by eaves and cornices as well as green painted ironwork.

During the British period a new construction technique arrived in Chamba
from Kashmir – *dhajji dewari* – a building technique influenced by Persian and
European precedents, and a refined version of the local *thatara tola* (Hicyilmaz
et al. 2011). This style was preferred by affluent families in the 19th century
for its less rustic and "neater" appearance distinguished by white plastered
walls, semi-circular arches, cornices, sloping roofs, wooden eaves, and deeper
verandas. Eventually a new vernacular form emerged in the region that integrated *dhajji dewari* with *thatara tola*. Such intersections of building traditions
are reflected at several sites in town; sometimes the lower level of structure
exhibits *thatara tola* and upper floor *dhajji dewari*, or vice versa.

Ongoing threats to cultural resources

Natural hazards

Much of the Himalayan region remains prone to natural hazards that pose
risk to life and safety of residents, and threaten the local economy and cultural resources (Al-Nammari & Lindell 2009; Al-Nammari & Alzaghal
2015). Chief among these are earthquakes, fires, cloudburst, flash flooding,
and landslides, which have increased in recent decades due to unchecked

growth and climate change. The district of Chamba (including Chamba town) is a high intensity earthquake zone (categorized seismic zone IV and V as per the seismic zoning map of India). The most recent earthquake to hit Chamba, on 24 March 1995, wreaked havoc on the town's built fabric, including historic buildings. About 70% of the built stock suffered cracks and fissures of varying intensity. Some temples tilted up to 20 degrees from their central axis with supporting columns displaced. Additionally, they were left with surface and structural cracks and warped stones (Joshi & Thakur 2016). The vernacular construction in town has evolved to mitigate seismic impacts; both *thatara tola* and *dhajji dewari* construction allow for "breathing space" and flexibility, which help withstand strong tremors. It is not surprising, then, that these buildings performed relatively better during the 1995 earthquake than the newer construction (Joshi & Thakur 2016; Town and Country Planning Department H.P.). Even today, the new construction in town does not exhibit seismic preparedness.

Fires have also long posed safety risks to the residents of Chamba and impacted the built environment. The risk is especially acute in the town since historic vernacular structures often include the traditional material of wood, and the built form is compact, with limited fire escape routes. In 1937, a large fire broke out in town that razed the elite *Chountra Mohallah*, after which fire hydrants were installed and a reservoir constructed for piped delivery of water to improve the town's firefighting capacity ("District council seeks" 2017). However, that system has since been stressed with rusted pipes and demand far exceeding capacity. More recently the town has been investing in installation of additional fire hydrants. The public works and irrigation departments are now coordinating to ensure their continued maintenance and proper use.

The cultural landscapes of Chamba Valley are further prone to flooding and landslides that are routine occurrences in the region. The frequency of landslides has increased in the past few decades due to deforestation, and road and other construction activities. Additionally, the upper reaches of the valley experience flash flooding due to cloudbursts, glacial lake outbursts, and temporary blockages of the river channels. In addition to life and safety risks to residents these natural disasters threaten the sacred landscapes of the region, particularly the traditional water systems such as *kuhls*, which are already stressed by the altering natural environment. With streams drying, some of the water fountains are currently obsolete. Climate change appears to be increasing the combined frequencies of hazards in the region. For example, in the last two years, the Chamba District experienced flash flooding and landslides in August 2019 (Shri Puri 2019); a magnitude 3.6 earthquake in February 2020 ("Moderate intensity earthquake" 2020); and heavy rains and flash flooding in June 2020 ("21 killed after heavy rain in Himachal" 2020). Among other issues, such developments sever community ties to the sacred landscapes as they are unable to carry on their everyday or traditional practices in any meaningful way.

Unplanned growth development

Even as the historic core of Chamba is notified as "Heritage-Cum-Conservation Zone", rapid growth and development in town poses a great risk to its vernacular built environments (and sacred cultural landscapes outside of town). To circumvent bureaucratic hassles of obtaining a demolition permit, typically a historic vernacular structure is first pulled down before filing a permit application for the new construction. Most additions, alterations, and new construction flout the design regulations of the "Heritage-Cum-Conservation Zone" by using modern construction materials and techniques. With improved transportation linkages with larger urban centres of the country, a range of modern building materials are readily available in Chamba at competitive prices. Furthermore, the traditional construction materials, particularly timber, are expensive and difficult to obtain due to restrictions on mining and lumbering. As a result, traditional knowledge on building crafts is fast dwindling, and the historic feel of the town is being compromised.

The courtyards in the traditional *shala*-homes are now covered with newer structures. This is especially evident in the elite *mohallahs* where the plot sizes are larger and can accommodate new homes or extensions to existing homes. Even with this trend, the elite *mohallahs* generally retain a greater range of vernacular homes compared to the non-elite *mohallahs* with smaller plots, where limited room for expansion fuels the demolition of older homes in favour of taller modern structures. When the traditional structure is retained, often load-bearing columns support modern cantilevered upper floors. The open spaces in *mohallahs* which once served as community spaces have been encroached upon by residents for informal commerce and by the town for surface parking. Infrastructure improvements in town have furthered the loss of vernacular fabric; street widening and parking projects routinely relied on tearing down existing vernacular stock. Pitched roofs, which are a character-defining feature of hill architecture in India, are giving way to flat roofs. The public spaces have also been compromised; for instance, the colonnaded arcade of the bazaar along the *Chowghan* has been altered beyond recognition with additions and alterations by the shopkeepers.

Even the protected or maintained structures have their own challenges. Some community temples have been painted with acrylic paints; others have lost their orchards or groves to new construction. Many additions and alterations to historic temples use modern materials such as concrete, cement plaster, and acrylic paint, which take away from the historic feel even as some of these changes may be reversible. One temple has been entirely enclosed within a concrete structure and is unrecognizable.

Figure 9.4 View from the interior courtyard of Rang Mahal shows the building in a poor state of repair due to lack of maintenance and incompatible uses. Source: Sakriti Vishwakarma

The Akhand Chandi Palace complex has been in use as an educational institution for several years now, which has had its toll on the historic structures. Signs of deferred maintenance and patchy repairs are evident. The open spaces of the palace property have been encroached by newer structures. Currently the Rang Mahal[6] is in use as a warehouse and training centre for local handicraft and for state administrative offices related to handloom and handicrafts. While adaptive reuse is generally a good approach to maintaining historic buildings, the new uses need to be compatible (Figure 9.4). Rang Mahal shows signs of distress as manufacturing uses have contributed to its degradation. (One of its staircases is on the verge of collapse.) Darbar Hall is used as an assembly hall and for public functions of the Government College. Even as it retains its historic character, deferred maintenance has taken its toll; currently, access to the second floor of the Darbar Hall has been closed off to the public due to safety concerns.

Current status of cultural resource protection

As elsewhere in India, in Chamba the responsibility of maintaining historic sites falls under three levels of government: national (through the Archaeological Survey of India, or ASI), state (Himachal Pradesh State Archaeology Department), and municipal (Municipal Council Chamba). The ASI maintains six temple sites (including the Lakshmi Narayan group of temples) and one rock sculpture site. The ASI designation ensures preservation and maintenance of the site under the Ancient Monuments and Archaeological Sites and Remains (Amendment and Validation) Act 2010. Also, a 100-metre no-construction zone and 200-metre regulatory zone around the protected monument preserve the site's context. However, in Chamba, enforcement of these rules is uneven and the contexts of the ASI sites have been impaired to varying degrees. The agency purposefully does not interfere in the way the temples are managed by the religious bodies, which includes routine maintenance, visitor management, and provisions for infrastructure. While understandable from the perspective of ensuring religious independence from government, often this lack of oversight and consultation has a negative impact on the built fabric.

At the state level, the Himachal Pradesh State Archaeology Department maintains one site: Chamunda Temple. Additionally, the Town and Country Planning Department of Himachal Pradesh established an advisory committee based in Shimla – "Heritage Conservation Advisory Committee" – that monitors areas of historical significance, particularly sites that possess distinct architectural styles important for tourism state-wide. State-wide, such areas have been designated as "Heritage-Cum-Conservation Zones" where guidelines impose regulations and approval on repairs, construction, demolition, and signage (Town and Country Planning Department H.P.). The historic core of Chamba was notified as a "Heritage-Cum-Conservation Zone" in 1995 to maintain its historic character by regulating new developments through design controls. To ensure controls on façade, height, materials, and massing, approval is required for architectural drawings for new projects and alterations (Town and Country Planning Department H.P.). However, on-site explorations of Chamba paint a different picture. The historic vernacular built form in Chamba that gives the town its distinct image is rapidly transforming, deteriorating, and disappearing due to incompatible new construction and incongruous additions and alterations. Vernacular construction in the region exhibits better adaptability than more recent imports to natural conditions including circulation, light, and insulation as well as seismic preparedness. The sloping roof does not accumulate snow, and the walls filled with stone rubble provide better insulation from harsh weather and allow for shock absorptions during an earthquake. Traditional temples in Chamba have a broad base with tapering tops so they can withstand seismic shocks up to a certain degree of ground shaking. However, if they are not seismically retrofitted, they remain vulnerable during a higher impact earthquake. Far more seismically risky, however, is the newer construction of brick and concrete in the *mohallahs*, especially when they are poorly constructed without expert supervision and proper reinforcements. This makes their presence hazardous to the life and safety of residents, as well as to traditional buildings in the vicinity, as collectively the neighbourhood is likely to perform poorly during a seismic event (Figure 9.5).

Like the vernacular heritage stock, the cultural and sacred landscapes of Chamba remain unprotected and prone to development, encroachment,

Figure 9.5 View of an *ekshala* dwelling in Surara Moholla with a modern, brick and concrete extension at right angle, to the left of the image. The debris in the foreground is the construction material for a demolished *thatara tola* construction that the new extension has replaced. Also seen along with the debris is galvanized roofing sheets that are fast replacing the traditional timber roofs with slate shingles. Source: Sakriti Vishwakarma

road building, and mechanized agriculture. Their neglect, combined with risks associated with the many forms of natural hazards, impact both the natural systems and the cultural and sacred values associated with these landscapes. Any protection measures for historic properties must take into consideration the seismic realities of the land, and invest in the maintenance of traditional building techniques that are better suited for both the climate and seismic threats to the region. Additionally, building permits should require new construction to exhibit seismic preparedness and contextual appropriateness.

Going forward

Our work demonstrates that Chamba is rich in both built heritage and sacred landscapes, with which both residents and visitors connect. These enrich everyday life by offering meaning and connections. However, the town's heritage remains largely unprotected outside of a select set of monuments. One clear, crucial step in rectifying these issues going forward is the development and enforcement of better oversight. The regulations of the "Heritage-Cum-Conservation Zone" in Chamba need to be better enforced and integrate seismic retrofitting in its permitting process. Additionally, the State Disaster Management Plan should emphasise mitigation, rebuilding, and risk reduction for historic structures, which it currently does not. For such changes to be successful, however, there must also be a fundamental shift towards understanding the entirety of the built and intangible heritage as an interwoven cultural product that reflects age-old wisdom and practices. Such traditional knowledge systems prevail in Chamba in the form of vernacular architecture, historic irrigation and water systems, sacred groves and vegetation, common traditions and cultural practices. It is vital that the community and leaders of Chamba, as well as conservation professionals at the state and central levels, recognize the unique value of Chamba's rich and multifaceted heritage now, before the chance to maintain its historic integrity is lost forever.

Notes

1 For more on temples in Chamba particularly their typology see Potdar et al (2017) titled "Nature, Culture and Humans: Patterns and Effects of Urbanization in Lesser Himalayan Mountainous Historic Urban Landscape of Chamba, India" in the *Journal of Heritage Management*.
2 This bridge from 1895 replaced an older wired suspension bridge that was destroyed by a flooding in 1894.
3 British colonial influences include the use of dressed and plastered stone and brick masonry in the structures, and semi-circular arches with stained glass windows.
4 Information on the town's morphology has been pieced together by consulting among other sources an unpublished studio project from 2014 in the Master of Architectural Conservation program in the School of Planning and Architecture, New Delhi. The co-author of this chapter participated in that studio as a student and her team including Shobana Devi, and Bharti Sikri complied data on the architecture of the town, which has informed this section in the chapter.
5 For more on the topic of antiquities in the Chamba region refer to Vogel, Jean Philippe, two volumes on the *Antiquities of Chamba State*, 1911.
6 Rang Mahal, built in the 18th century, exhibits a hybrid architectural style showcasing English turrets and Mughal multifold arches.

References

21 killed after heavy rain in Himachal; schools to be shut tomorrow (2020, June 16) *Hindustan Times*. Retrieved from: https://www.hindustantimes.com/india-new s/10-killed-after-heavy-rains-lead-to-flash-floods-landslides-in-himachal-prades h/story-wE4EEuzkxK3lqp6w379nqI.html

Al-Nammari, F. M., & Lindell, M. K. (2009). Earthquake recovery of historic buildings: Exploring cost and time needs. *Disasters*, *33*(3), 457–481.

Al-Nammari, F., & Alzaghal, M. (2015). Towards Local Disaster Risk Reduction in Developing Countries: Challenges from Jordan. *International Journal of Disaster Risk Reduction*, *12*, 34–41.

Anthwal, A., Gupta, N., Sharma, A., Anthwal, S., & Kim, K. (2010). Conserving biodiversity through traditional beliefs in sacred groves in Uttarakhand Himalaya, India. *Resources, Conservation & Recycling*, *54*(11), 962–971. doi:10.1016/j. resconrec.2010.02.003

Baker, J. (1997). Common Property Resource Theory and the Kuhl Irrigation Systems of Himachal Pradesh, India. *Human Organization, 56*(2), 199–208.

Bandarin, F., Hosagrahar, J., & Albernaz, F. S. (2011). Why development needs culture. *Journal of Cultural Heritage Management and Sustainable Development, 1*(1), 15–25.

Bharti, K. R. (2002). *Chamba Himalaya: Amazing Land, Unique Culture.* Indus Publishing Company.

Bhatnagar, M. (2008). Chamba: Urban evolution of an ancient town in the Himalaya. *Journal of the Development and the Research Organization for Nature, Arts and Heritage,* V(1), 39.

Charak, S. D. S. (1978). *Himachal Pradesh* (1st ed. ed.). New Delhi: New Delhi: Light & Life Publishers.

Dave, B., Thakkar, J., & Shah, M. (2012). Details of Resistance: Indigenous Construction Systems in Himachal Pradesh. *Context, 9*(1), 4–17.

Directorate of Census Operations (H.P.). (2011). *District census handbook-Chamba* (Series-handa03 Part XII-B ed.)

District council seeks 3 more fire hydrants (2017, May 25), *The Tribune.* Retrieved from: https://www.tribuneindia.com/news/archive/himachal/district-council-se eks-3-more-fire-hydrants-412177

Eck, D. L. (1993). *Encountering God: A Spiritual Journey from Bozeman to Banaras.* Boston: Boston : Beacon Press.

Goetz, H. (1955). *The Early Wooden Temples of Chamba.* ("Instituut Kern," Leyden. Memoirs). Leiden: E.J. Brill.

Hāṇḍā, O. (2001). *Temple Architecture of the Western Himalaya: Wooden Temples.* New Delhi: Indus Publishing Co.

Hicyilmaz, K., Bothara, J. K., & Stephenson, M. (2011). *Dhajji dewari.* (Housing report No. 146).World Housing Encyclopedia. Retrieved from http://db.world -housing.net/building/146/

Hosagrahar, J. (2001). Mansions to Margins: Modernity and the Domestic Landscapes of Historic Delhi, 1847–1910. *Journal of the Society of Architectural Historians, 60*(1), 26–45.

Hosagrahar, J. (2002). South Asia: Looking Back, Moving Ahead-History and Modernization. *Journal of the Society of Architectural Historians, 61*(3), 355–369.

Hutchison, J., & Vogel, J. Ph. (1982). *History of the Panjab hill states.* Simla: Dept. of Languages and Culture, Himachal Pradesh. (Originally published in Lahore: Supt., Govt. Print., Punjab, 1933).

Hutchison, J. (1904). *Gazetteer of the Chamba State.* Indus Publishing.

Institute of Integrated Rural Development, Shimla. (2011). *Stock Taking Report - Integrated District Planning Exercise, District Chamba, Himachal Pradesh.*

Joshi, M., & Thakur, V. (2016). Signatures of 1905 Kangra and 1555 Kashmir Earthquakes in Medieval Period Temples of Chamba Region, Northwest Himalaya. *Seismological Research Letters; Seismol.Res.Lett., 87*(5), 1150–1160.

Kaul Deambi, B. (1985). *History and culture of ancient Gandhāra and western Himalayas: From Sāradā epigraphic sources.* New Delhi: Ariana Publishing Group House.

Khandalavala, K. (1989). The Princess's Choice. In Ohri, V. C., Khanna, A. N., & Himachal Pradesh (India) Department of Languages (Eds.), *History and Culture* of the Chamba State, a Western Himalayan Kingdom: Collected papers of the seminar held at Chamba in 1983, 1–14. New Delhi: Books & Books.

McKay, A. (2015). *Kailas histories: renunciate traditions and the construction of Himalayan sacred geography*. Brill, Leiden; Boston.

Mehta, M. (2011). *The Mouse Who Would be King: Innovating Tradition in the State of Chamba*. ProQuest Dissertations Publishing.

Meister, M. W. (1979). Maṇḍala and Practice in Nāgara Architecture in North India. *Journal of the American Oriental Society*, 99(2), 204–219. doi:10.2307/602657

Moderate intensity earthquake hits Himachal's Chamba district (2020, February 26), *India Today*. Retrieved from: https://www.indiatoday.in/india/story/moderate-intensity-earthquake-hits-himachal-chamba-district-1650077-2020-02-26

Negi, T. (1963). *Himachal Pradesh district gazetteers*. (Gazetteer of India). Batala: Standard Print. Press.

Potdar, K., Namrata, N., & Sami, A. (2017). Nature, Culture and Humans: Patterns and Effects of Urbanization in Lesser Himalayan Mountainous Historic Urban Landscape of Chamba, India. *Journal of Heritage Management*, 2(2), 169–188.

Rahul, A. Sood, A., Singh, Y., & Lang, D. (2013). *Thathara houses in Himachal Pradesh*. (Housing report No. 170).World Housing Encyclopedia. Retrieved from http://db.world-housing.net/building/170/

Shashi, S. S. (1971). *Himachal-nature's peaceful paradise*. Indian School Supply Depot, New Delhi.

Shri Puri. (2019). Heavy rains trigger flash floods, block roads in parts of Himachal Pradesh (August 18, 2019). *The Times of India*. Retrieved from https://timesofindia.indiatimes.com/city/shimla/heavy-rains-trigger-flash-floods-block-roads-in-parts-of-himachal-pradesh/articleshow/70711644.cms

Sinha, A. (2006). *Landscapes in India: Forms And Meanings*. Boulder: University Press of Colorado.

Sinha, A. (2014). The Sacred Landscape of Braj, India. *Landscape Journal*, 33(1), 59–76.

Town and Country Planning Department H.P. (a). *Heritage of Chamba Town*. Unpublished manuscript.

Part III

Emerging trends in heritage conservation

10 Making heritage accessible to all

Experiments with digital technologies for urban heritage conservation in India

Aishwarya Tipnis

Digital technologies for citizen engagement in heritage conservation

The new millennium brought about a tectonic shift in the understanding of heritage, from monumental edifices to non-monumental buildings of everyday use and their role in the life of the community. This coincided with rise of information technology that allowed for easy dissemination of information and enabled greater participation for content creation (Roued-Cunliffe and Copeland 2017, xv). It gave heritage enthusiasts a platform to engage and share information about a site or place on blogs and popular social media sites such as Facebook and Instagram. As philosophies of conservation evolve to become more people-centric, the potential of role digital technologies in citizen participation for heritage conservation is crucial (ICCROM 2015).

There has been considerable research on citizen heritage and methodologies for encouraging participation through digital technologies (Lewi and Smith 2016; Han et al 2014). Lewi and Smith have drawn parallels between digital tools for citizen engagement with citizen science, designed to bridge the gap between expert-driven scientific knowledge and the needs and concerns of the citizens. The main types of digital tools when co-related with citizen science can be divided into three types: curated sites (contributory), content hosting sites (collaborative), and social networking sites (co-created) (Lewi and Smith 2016, 2–6). It has been observed that most of the digital heritage projects in India have largely focused on creating a databank or online archive for heritage properties. These sites are usually built by institutions and disseminate information that is researched and curated by experts and fall within the category of contributory sites as defined by Lewi and Smith. These function as a web-museum of stories and narratives around artefacts or tangible cultural resources associated with heritage sites. As the focus shifts from narratives of authorised heritage discourses to stories of everyday people, the role of social media and other digital tools gain prominence. In the second category of co-creating, digital technologies have encouraged participation of citizens and enthusiasts in contributing to online heritage content they are familiar with. The third category of collaboration promotes citizen engagement through popular social networking sites like Facebook and messaging applications like WhatsApp. In India, heritage enthusiasts and activists use these platforms for active interactions and discussions on various heritage topics across the country.

This chapter discusses three projects by Aishwarya Tipnis Architects (ATA) that have employed the use of digital technologies with the aim of engaging diverse audiences in heritage conservation. The first project is a private restoration case from Delhi, where ATA developed a blog and employed social media tools for disseminating information about the project. It also became a medium for educational workshops and enabled the team to share the process of restoration with a wider audience. Our objective was to inspire and empower residents to consider restoring their heritage properties in place of demolition. Our second project is a cross-disciplinary digital humanities project that aims to define the heritage of a place through documentation of tangible and intangible heritage. As part of the project, ATA developed a website to spread awareness and disseminate academic research generated through the project. The third example is a self-initiated collaborative citizen engagement and heritage conservation project that relies on an interactive web-based platform for creating a "web-home": a repository of oral histories, memories, and stories about a place. All three projects employ overlapping but unique strategies that offer lessons for practice in the use of digital technologies for participatory heritage conservation in urban India.

The Hero case study: restoration of Seth Ram Lal Khemka Haveli, Kashmere Gate, Delhi

Inspiring change through digital tools: blogging and social media

This project, which began as an interior design exercise, has proven that a small project can have a wide-ranging impact for the regeneration of an area. The site is a private *haveli* that was home to a Marwari joint family, nestled within the bustling car market in Chotta Bazaar in the Kashmere Gate area. Partly derelict, the owner had consolidated the ground and mezzanine floor of the *haveli* after buying out his cousins in 2010. Being a traditional family, the owner wanted to renovate the house so that he could seek arranged marriages for his three sons. He was unaware of the heritage value of his building or what listing of a heritage property entailed. For the client, the ancestral nature of the home was paramount; more than the love for the historic fabric, it was the associated memories that he valued and continued to reside here even as members of his extended family had moved out.

The first site visit in the summer of 2010 revealed that the *haveli* was in a high degree of distress with severe structural cracks, as well as rising dampness to almost 8'0" on all the walls. While it had not undergone any repairs for over 40 years, the family continued to live in a section of it. Our first question to the owner was if he had considered restoring the property. Baffled at my question, he reiterated that it was his home and not a 'museum'. Thus, began a long interaction that would ultimately lead to education of the homeowner about urban heritage (particularly his building in the context of the walled city) and its continuing contemporary relevance. There were many lessons learnt from this restoration project in the walled city that lasted over 7.5 years from conceptualization to completion.

In a scenario where a penny saved was a penny earned, the onus of actualisation and implementation of this project fell on the professionals. The project demanded the conservation expert to not just be a restoration architect, but also a researcher, project manager, and a historian. This project was designed to be a collaborative one, where the decision-making process involved lengthy discussions and sometimes negotiations between the architect, craftsmen, and the client. Discourses ranged from convincing the client to restore his *haveli* in an appropriate way, to the choice of materials and finishes within budgetary constraints. Such interactions, which were sometimes fraught with disagreements and arguments, are what led to the development of a methodology that involved interactions and research that helped convince the client about the value of his property. Unlike in the West, where restoration is a more specialised industry and products such as lime mortars, heritage tiles, lime-wash, etc., are readily available off the shelf, this project required constant improvisations necessitating additional research and analysis to find the appropriate materials for restoration. The internet-savvy client, in an effort to be more informed, conducted his own research and shared with the team ideas on restoration techniques, new materials, as well as furniture and finishes within his budget that would complement the historic interiors.

One of the most significant breakthroughs on this project was the use of lime mortar in the process of restoration. The benefits of lime mortar are well known in professional and academic literature – practical information on where and how to procure it, however, was missing. The project demanded significant research on how to re-create historic lime mortar, including interviews with experts, professional peers, as well as multiple craftsmen. It became clear that there was no one formula for lime mortar: every craftsman had his own vernacular version. We decided to try the various methods we came across, and made multiple samples to see which best suited the site and then decided the final proportions. It took multiple trials and significant experimentation to get a mix that worked for the project. The team also had the daunting task of developing the right apparatus to make the mortar. We also faced the challenge of locating a large enough open space in the walled city that could accommodate a conventional mortar mill hurled by a bullock, tractor, or auto rickshaw. Our other option included installing a conventional mill somewhere on the outskirts of the city, but the narrow bylanes of the walled city with restricted traffic movement for commercial vehicles made this option impractical. Our work was further complicated by the family being in residence throughout the project.

Our team was able to innovate within the modest means available, using a second-hand mortar and grinder wheels from an *idli* mixer (kitchen gadget) to form an assembly for preparing the mortar mill in the most cost-effective manner. This home-made lime mortar was then used for the entire restoration process.

Sharing and learning from each other has always been the epitome of our work and we inculcated this into this project from the very beginning. Academic curriculum for architecture in India rarely includes the study of traditional materials, especially lime, so we were keen to share our experiences with lime mortar with students and young professionals. Thus was conceived what we called "The Lime Workshop". With a Facebook page and online posts, we received 120 applications from enthusiastic students to the extent that refusals had to be sent owing to a space constraint on site. We conducted three back-to-back workshops in March 2014 in batches of 30 students. The workshop comprised a lecture and practical demonstration of hands-on application of lime mortar. We were humbled by the overwhelming interest in lime mortar as a material and techniques associated with its application and fielded requests to conduct more workshops (Figure 10.1).

The success of this workshop prompted the creation of a blog titled "The Haveli Project: A Common Man's Guide to Restoring A Heritage Home" [www.thehaveliproject.blogspot.in]. It was meticulously and periodically updated with the goings on the site and we shared lessons from the work

Figure 10.1 Lime workshop in progress at the Seth Ram Lal Khemka Haveli.
Source: Aishwarya Tipnis Architects

being conducted on a regular basis. We also encouraged students to visit the ongoing site works to learn from our experience. Students from the Apeejay School of Architecture and Planning, Greater Noida, the University of Cologne, Germany, and students from the Departments of Urban Design and Architectural Conservation from the School of Planning and Architecture in New Delhi visited the site in groups through various stages of the project. By this time, our client had become comfortable to use lime himself; so much so that when students visited, he would share his own experiences with them. As a practitioner, this was the real success of the project where the originally hesitant client became a champion of a historic and sustainable material.

As noted by Verma (2014), the state government was keen to adopt a tourism led approach for the walled city and was providing incentives to *haveli* owners for converting their premises into guest houses or art galleries. Our project demonstrated that *havelis* were capable of becoming comfortable contemporary dwellings as well. The blog became the medium for sharing the story of a *haveli* owner's journey through the restoration process and it piqued the interest of many journalists interested in the regeneration of the walled city. Various stages of the project were featured in national and international papers that brought attention to the *havelis* in Old Delhi. As noted by Vergese (2015), our work spurred interests in other such rehabilitation and restoration projects of *havelis* within the walled city. This work was featured by the ICOMOS Evaluation Committee as a private citizen's initiative within the walled city. Further, it led to INTACH Delhi commissioning us to write a *Manual for Owners and Occupiers of Havelis in Shahjahanabad for Restoring Historic Properties* based on our experience.

The role of the conservation professional therefore in this case became that of an enabler that empowered the owner of a heritage property to restore his own property, guiding him to adopt the appropriate methodology and materials for conservation. Additionally, creating a public blog to share the project and its methodology went a long way in propagating the value of conservation in keeping an ancestral home alive. The use of digital technology therefore helped in creating awareness, connecting people, and making heritage and conservation accessible to many. The blog emerged as a space for professionals to share their work in an open environment and learn from each other's experiences. The project was therefore called the Hero case study; to underscore the role an individual can play in regeneration efforts. It is clear that if individual home owners can take responsibility for restoring their historic properties then we can achieve large-scale regeneration without relying on government-led initiatives.

Engaging with the digital world: experiments with the Dutch in Chinsurah

In 2013, the Embassy of the Kingdom of the Netherlands commissioned ATA to develop a comprehensive listing and mapping of what remained of the

Dutch built heritage in Chinsurah. Chinsurah was an erstwhile Dutch trading post along the River Hooghly in West Bengal, which gained prominence between the 17th and 19th centuries. Like the other European trading posts of Serampore (Danish), Chandernagore (French), and Bandel (Portuguese), Chinsurah moved into British hands in 1840 (Crawford 1902).

This region has been recognised as a unique cultural landscape embodying immense cultural value and having the potential of being nominated as a world heritage site. The Government of West Bengal has been keen to develop this region as a tourist zone and efforts are being made by multiple agencies in this regard (*Indian Express*, 2015). Today, these erstwhile trading posts have grown into small towns and are part of metropolitan Kolkata (Calcutta). The local community comprises of educated citizens mostly engaged in service who travel daily on the suburban trains to Kolkata for their jobs. These towns are now lucrative as they offer affordable housing for daily commuters and thus find themselves at the centre of developer-led, project-driven urban development. The cultural landscape once dotted by palm trees and ground plus one-storeyed spacious mansions or town houses is rapidly being replaced by piecemeal developments of multi-storey apartments. The layered history of these towns and their urban fabric seems to be inconsequential to this new wave of urban development that threatens to make these suburban towns generic clones of each other and rob them of their individual identity and sense of place.

As part of the "Shared Built Heritage" [www.culturalheritageconnections.org] programme of the Netherlands, the project evolved from architectural mapping to a cross-disciplinary digital humanities project in collaboration with Presidency University, Kolkata. This was then presented via a new website [www.dutchinchinsurah.in] in the form of an interactive timeline. When history relied on narratives rather than the built fabric on site, it became even more essential to covey that story accurately. Our historical research led us to maps of Chinsurah in the archives in the Netherlands as well as in Kolkata, which helped the team to understand the location of Fort Gustavus[1], as well as decipher the urban morphology of the town. Digital tools came to our rescue – we were able to overlay the historical maps using open source Google technology. This was then presented as a section on the website to disseminate information on the historical evolution of the town. While conducting the architectural mapping of the town, the team also realised that most locals were not aware of its history and heritage. There were many narratives of the rich merchant families that had built the town, but these oral histories needed to find a place in the official discourse of the history of this town.

Our team from ATA decided to engage with the local community, particularly the youth, to understand their perspective of how to best interpret the town's history. The role of the conservation professional therefore became that of a design and community outreach strategist. In keeping with the low budget of the project, the engagement was designed as competitions and workshops on painting, comic book, and photography. This community engagement led to the creation of a children's storybook (available for download from the website in both English and Bengali) featuring the Dutch in Chinsurah. It was illustrated with the drawings created by the local school children during organized competitions and workshops. Additionally, based on community engagement, we generated a heritage walk route and map (available as a free download) and trained some local students to guide this walk. Architectural mapping was presented in the form of geo-referenced Google maps, with information about the listed buildings, their photographs, etc., included as part of the website. A similar section of the website has been created for the Dutch Cemetery, describing each of the graves, their tombstones, and information on people buried there. The Dutch in Chinsurah website was further supported by a Facebook page, wherein people could share their comments and ideas.

The project, conceptualised in collaboration with Presidency University, as well as the Embassy of the Kingdom of the Netherlands in India, was designed as an online archive, a digital museum for disseminating information curated by experts for visitors. The local citizens were involved in a moderated manner for data generation such as the comic book, heritage walks, or storybook projects which were again designed and curated by technical experts. The project led to creating awareness about the built heritage of Chinsurah. Other professionals and enthusiasts used the site as a platform to conduct parallel activities; two Dutch citizens living in West Bengal developed an enterprise around cycling tours which included some of the landmarks of Chinsurah. Two publications about the Dutch Heritage in Chinsurah were launched in 2014. Additionally, as a direct take-away from the project, the Embassy of the Kingdom of Netherlands executed an interpretation and signage project, building plaques and information signage for the heritage sites, which were then installed in Chinsurah in January 2017 (Figure 10.2). We hope that the enthusiasm and awareness created about the heritage in Chinsurah through this project in future translates into physical restoration of some of the landmark buildings and public spaces within the town.

Collaborative mapping: the heritage and people of the Chandernagore Project

Chandernagore is the erstwhile French settlement on the banks of the River Hooghly within the agglomeration of the other trading posts of the European nations. It was one of the five French trading posts in India, the others being Pondicherry, Mahe, Yanam, and Kariakal. However, Chandernagore remained in French occupation until 1948, when its citizens voted to be part of India, while the other settlements joined the Republic of India six years later in 1954. In the decades following independence, with a decline in trade and other economic activity within the town,

Figure 10.2 Signage and interpretation plaques installed at Chinsurah West Bengal.
Source: Aishwarya Tipnis Architects

Chandernagore was relegated to a sleepy suburban town in close proximity to Calcutta. Interest in French heritage in India grew in the 1980s, with most of the efforts focussing on Pondicherry. In 2010, a Paris-based NGO *Vieilles Maisons Françaises*, in collaboration with the Embassy of France in India and French Heritage in India society, expressed interest in exploring the heritage of Chandernagore. ATA was commissioned to prepare an architectural inventory and mapping exercise for identification of French heritage in Chandernagore. The outcome of the project was a listing of 99 buildings of Indo-French heritage value in 2011. However, due to the lack of legislative protection, the buildings were under severe threat of redevelopment. The research also highlighted that one of the main challenges of the preservation of French heritage in Chandernagore was the lack of awareness and appreciation of this heritage within the community. Therefore, in 2015, ATA launched a digital project called Heritage and People of Chanderagore [www.heritagechandernagore.com], supported by

Vieilles Maisons Françaises. The project applied the lessons learnt from the Dutch in Chinsurah project, which showed that the website was static and did not allow adding or updating information after the project's execution. The architectural mapping of the town of Chandernagore had highlighted that heritage was not just about the buildings, but also about the people who lived there then and who live there now.

The Heritage and People of Chandernagore Project was therefore designed as a collaborative mapping project, with the objective of not only identifying the built heritage of the town but also the intangible heritage and what the local citizen perceived as their heritage. A call for applications was placed on social media for "citizen historians", interested in volunteering and collaborating on co-creating a historical narrative for Chandernagore. Many students from the town and nearby areas, particularly students of history from Chandernagore College, Presidency University, as well as Jawaharlal Nehru University, New Delhi, joined the project. Their task was to go from door to door, collecting oral histories and narratives, sometimes captured on video and sometimes in written format. The citizen historians then added their blurb on the project blog, which was widely shared on social media, attracting many comments and suggestions (Figure 10.3).

The historical overview of the town was created through a graphic timeline. Maps from different historical periods were digitally overlaid to allow

Figure 10.3 Citizen Historians collect oral history at Chandernagore. Source: Aishwarya Tipnis Architects

visitors to develop their own understanding of the historical evolution of the town. Furthermore, the website was designed so it could gather information via crowdsourcing where users could post content pinned on geolocations of places connected to memories or associations. While conducting the survey, it became evident that the majority of the populations in the town were not digital savvy and did not even own smartphones. The strategy was therefore modified from being individual dependent to that of being technically assisted by citizen historians who were encouraged to collect memories and use their infrastructure (smartphones or desktops) to upload the content on behalf of the community. The citizen historians also co-created a heritage trail through the town and developed their own map which was shared through the website.

The third aspect of the project was that of citizen engagement through workshops and competitions. An important lesson from the Dutch in Chinsurah project was that the local community was not responsive to the workshop model but far more enthusiastic about competitions. Therefore, various kinds of competitive engagements were designed around the theme of "Heritage and People of Chandernagore". These included painting, creative writing, quizzes, as well as games (Brick by Brick Lego) and a Franco-Bangla cuisine competition that helped develop camaraderie among townsfolk around heritage matters. We were able to conduct two specialist workshops; the first was on photography, where the participants interacted with Sanjit Chowdhury, a well-known photographer who led them on a photo walk and shared photography techniques. Another workshop was conducted by illustration designer Chitra Chandrashekhar, where she inspired the participants to build their own stories around French heritage using a short comic strip format. Both the workshops generated interest within the student community but the number of participants was limited. Our objective was to remain inclusive and therefore our next activity was planned at the most prominent public space within the town, the riverside promenade popularly known as the Strand. The Strand is frequented by all citizens of the town, and on Independence Day weekend in 2015, a temporary stall was set up at the Strand. Passers-by and children were invited to come and engage with the citizen historians and volunteers. A cleverly designed treasure hunt was organized based on the heritage of Chandernagore, and schoolchildren were invited to come and play the game and learn about their town. Additionally, adults and senior citizens were engaged in conversations about what they liked about their town and what changes were desired. The conversations conducted by an urban designer, Sovan Saha, were graphically recorded on a large canvas on site by a community engagement specialist. This generated interest and curiosity among the passers-by and more people came to engage with the team on the future of their town. The project's presence on social media (Facebook and Instagram) was considerably popular with people from the community and those who had migrated abroad. This not only raised awareness about the town's heritage, but created a platform for the community to collaborate and generate historical contents supporting heritage conservation work. The project eventually received the support of the Embassy of France in India. This project has inspired other digital projects such as the Hugli River of Cultures Project [www.hugliriverofcultures.org] regarding the collation of oral history.

The primary objective of the project was to bring global attention to Chandernagore, this was covered by the local media and internationally shared through social media in France as well as India. In 2017, the Ambassador of France to India, His Excellency Alexandre Ziegler visited Chandernagore and the website became a tool to give him an orientation of the history as well as the heritage of the town. This was followed by a co-creation project between 32 students from India and France on a dilapidated historic building in town as part of the Bonjour India Festival in 2017. The students engaged with the local community to come up with ideas for the dilapidated building over seven days. This work was exhibited at the Strand (the main public space) of the town for the townspeople as well as the authorities. The interest generated by this led to the signing of an historic MoU between India and France for the restoration of French heritage along the Strand in Chandernagore.

Lessons for practice

All three projects employed different digital tools for the engagement of a new audience in heritage conservation. It was observed that unless the projects are initiated by the government or an institution, the sustainability of the project is a challenge once the initial funding is exhausted. This has also been observed in two very popular digital memory projects: the Indian Memory Project [www.indianmemoryproject.com] and 1947 Partition Archive [www.1947partitionarchive.org], both websites are now seeking financial support from visitors and patrons for continuing and taking the project to the next level. The issue of scalability and long-term sustainability has been addressed by the 1947 Partition Archive by not allowing free access to their collection and insisting on a paid subscription. Additionally, financial contributions and donations are also being sought from the public at large for setting up the world's largest partition archive at Amritsar.

Apart from the concerns about funding, experience of the Dutch in Chinsurah website indicated that digital heritage has to be more interactive and more collaborative in nature for it to succeed in the long term. We observed that as a static website, offering a curated content, it generated short-lived enthusiasm in the community; the success of the website was largely as a repository of academic information. In the case of the Heritage and People of Chandernagore Project, since the nature of the project was more collaborative and there was a direct engagement of the community in co-creating some of the content, the enthusiasm lasted for a couple of months. However, once the euphoria subsided and the citizen historians moved on to other projects, the traffic on both social media and the web-site dwindled. Both these examples highlight that the primary issue is that of longevity, particularly the long-term engagement and enthusiasm of the community. The websites and social media accounts for both projects are personally maintained by ATA even after the projects ended several years ago; it is clear that without ongoing resources to maintain content and participation, the sites will not be sustainable.

The second significant lesson from both Chandernagore and Chinsurah projects has been the need to integrate digital projects with physical projects for active engagement of the community in heritage conservation. The Dutch in Chinsurah website was launched in April 2015, the project was followed up with a seminar and panel discussion on the subject of sustainable tourism in the region by UNESCO in October 2015. The publications of the research as well as installation of interpretation panels for key heritage sites in January 2017 further reinforced the identity of the town as a place of shared cultural heritage. In the case of Chandernagore, while the website was launched in February 2016, events such as the visit of the Ambassador of France and the announcement of the Bonjour India Project in February 2017 re-activated the Facebook page of the project and volunteers from the town expressed their interest in participating in the next phase of the project. This is also observed in the case of the 1947 Partition Archive [www.1947partitionarchive.org], where the content of the research has now led to the setting up of a formal Partition Museum at the restored Town Hall at Amritsar. Promotional events such as a pop-up museum to commemorate 70 years of Partition have been organised at the Partition Museum [Remembering Partition Museum of Memories 4th–7th August 2017].

Another significant lesson from both projects has been the need to first assess the capacity of the community regarding their familiarity with the technology used; this will enable better decision-making in adoption of relevant technology for the project. Heritage projects based on digital platforms have to work on the model of cross-disciplinary collaboration between conservation professionals, website designers, and computer technicians. In both the Chinsurah and Chandernagore projects, it was a strategic decision to develop a mobile-friendly or responsive website rather than a mobile application. The main purpose of both the websites was dissemination of information and the content was to be curated within the framework decided by the technical experts on the project. Similarly, once the restoration of the Seth Ram Lal Khemka Haveli officially wrapped up in 2016, the blog remained as an online archive of the project which is periodically updated and continues to disseminate information on the project.

In conclusion, digital technologies have not only enabled easy access to information but also provided the platform with the means to empower citizens to participate in the protection of their town's heritage. Given the susceptibility of technology to constant change as well as the recurring costs towards the maintenance and upgradation of digital platforms, its posterity still remains debatable. Furthermore, when adopting a strategy for a digital heritage project, it is pertinent to understand the target audience, its needs and capacities to ensure that the project remains inclusive at all times. Several citizen historians trained in the Heritage & People of Chandernagore Project now have better employment prospects, adding to the project's sustainability.

These projects also showed that digital projects often tend to become elitist, working brilliantly in metropolitan regions but failing in semi-urban or rural settings simply due to the lack of hardware such as smartphones. Additionally, the projects showed that care needs to be taken to understand the level of comfort of a community with sharing information on public platforms or social media to ensure its long-term success. Finally, to achieve long-term sustainability, it becomes critical that all digital projects are supplemented with traditional physical conservation projects.

Note

1 Fort Gustavus, built by the Dutch as their main trading post was destroyed by the British in 1840.

References

Crawford Lt. Col. D. G. 1902. *A Brief History of the Hughli District.* Bengal Secretariat.

Han, Kyunsik, et al. 2014. Enhancing Community Awareness of and Participation in Local Heritage with a Mobile Application. *CSCW-Conference*, 2, 1144–1155.

ICCROM. 2015. Guidance: People *Centred Approached* to *Conservation* of Cultural *Heritage: Living Heritage.* www.iccrom.org/wp-content/uploads/PCA_Annexe-2.pdf [accessed 4 August 2017].

Indian Express. 2015. Article on "UNESCO: Need Public Participation to Preserve Hooghly's Heritage" on 2nd October 2015.

Lewi, Hannah and Wally Smith. 2016. Citizen Heritage: Provoking Participation in Place through Digital Technologies. *Historic Environment*, 28(2), 2–6. <search.informit.com.au/documentSummary;dn=445605951865450;res=IEL APA>ISSN: 0726-6715. [accessed 21 March 17].

Roued-Cunliffe, Henriette and Copeland Andrea, eds. 2017. Participatory Heritage. London: Facet Publishing, 173–185. https://books.google.co.in/books?id=TwkU DgAAQBAJ&pg=PA17&source=gbs_selected_pages&cad=2#v=onepage&q&f =false [cited 4 August 2017].

Union Government (Civil) Ministry of Culture. 2013. Report of the Controller and Auditor General of India, Performance Audit of Preservation and Conservation of Monuments & Antiquities. http://www.cag.gov.in/sites/default/files/audit_re port_files/Union_Performance_Ministry_Cultures_Monuments_Antiquities_1 8_2013.pdf

Vergese, Shiny. 2015. Bringing the Past Back: 165 Year Haveli Shows the Way. *Indian Express*. December 27.

Verma, Richi. 2014. Government May Help Restore Havelis in Delhi. Times of India, Delhi edition. December 16.

Web links

www.thehaveliproject.blogspot.in
www.dutchinchinsurah.in
www.heritagechandernagore.com
www.1947partitionarchive.org
www.indianmemoryproject.com
www.hugliriverofcultures.org

11 Reclaiming neighbourhood, rebonding community

Urban conservation initiatives for Kolkata's Chinatown

Kamalika Bose

Introduction

Historic preservation in the urban context becomes sustainable when it transcends the mere pickling of the past to play an active role in shaping the future of cities and local communities. Notions of heritage – both tangible and intangible – are strongly rooted in historic cores, not merely in the community's memory of the past but also in their active involvement in its contemporary relevance (Nagpal 2012). Kolkata has been slow (and often misguided) in addressing urgent preservation issues particularly those pertaining to its diverse cultural and ethnic heritage. Historically acknowledged as a melting pot of cultures and commerce, Kolkata's neighbourhoods are representative of a vibrant mix of people and communities. Some strands of its rich multicultural tapestry were shaped by immigrants of foreign origin in the 18th and 19th century. Some of these have faded while others are striving to retain their identity and maintain contemporary relevance. This chapter discusses the preservation and cultural continuity of one key community: the Indians of Chinese descent in Kolkata through the urban heritage in Chinatown – the only one of its kind in South Asia. It examines the role of community-based urban strategies, initiated under the Cha Project, in maintaining the continuity of this community's ethnic identity and heritage. The chapter attempts at disseminating the Cha Project's approach, methodologies and outcomes by extracting relevant lessons for preservation practice.

It is essential to address overarching but imperative questions on urban regeneration of ethnic communities. Who are the custodians of this built heritage in decline – the community itself, the city as a composite whole, or the local government? What is the value of shared cultural heritage and how can it be leveraged and revitalised? How can historic settlements of ethnic minorities be tapped as a unique resource in strengthening diversity of cultural experiences within Indian cities? How do such areas reconcile with change and continuity in their physical and cultural spheres while enhancing liveability and urban resilience for today?

Kolkata's cosmopolitan character is demonstrated in the multitude of architecturally diverse ethnic neighbourhoods, testimony to a rich mosaic of migrant communities. The Chinese in Kolkata are one such dwindling community with approximately 3,000 in population today (compared to the 4.5 million city population). The ethnic enclaves of Kolkata's former 'grey town,' largely comprising communities of foreign origin – including Armenians, Jews, Anglo-Indians, and Greeks – were separated from the Indian quarters. Many of these enclaves are no longer extant with the exception of the Chinese neighbourhood, which is languishing without support for preservation. The Cha Project is intended to serve as a catalyst for resurgence and urban revitalisation for this neighbourhood, which remains vulnerable to transformation (Figure 11.1).

What time is this place? Chinatown between nostalgia and reality

Chinatown, in central Kolkata, today spreads across an area of 1.5 sq. km, with Rabindra Sarani (formerly Chitpur Road) marking its western boundary at Poddar Court, and Chittaranjan Avenue (formerly Central Avenue) being the eastern access point. Six Chinese temples[1] built between 1850 and 1920, each representative of an ethnic Chinese sub-culture, are neighbourhood anchors that remain in use, although with a much-diminished congregation. The presence of *huigans* (or social clubs), merchant guild networks,

Figure 11.1 Down to a population of under 3000, the Indian Chinese community in Kolkata's Chinatown struggle to remain relevant within the city's broader narrative of development and change, while upholding their distinct cultural and economic practices. Source: Kamalika Bose

morning breakfast along Chattawalla Gullee (the southern edge of the neighbourhood) as a long-standing tradition, and the annual Chinese New Year festivities in Blackburn Lane reinforce *Cheenapara*'s socio-cultural vitality. More modest social institutions, artisanal workshops and provision stores, eating houses, and residential tenements, used and owned by the community, remain interspersed with redeveloped apartments, office high-rises, along with pockets of informal commerce. Chinatown today is a settlement with amorphous urban boundaries creating disparate architectural styles and mixed streetscapes, except in Blackburn Lane which retains its historic urban form. (Figure 11.2).

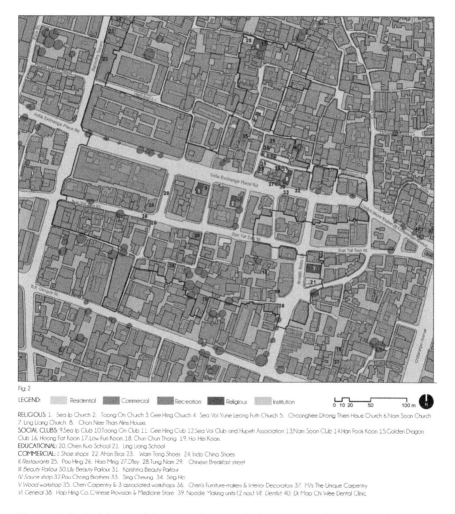

Fig. 2

LEGEND: ▨ Residential ▨ Commercial ▨ Recreation ▨ Religious ▨ Institution   0 10 20 50   100 m 🧭

RELIGIOUS: 1. Sea Ip Church 2. Toong On Church 3. Gee Hing Church 4. Sea Voi Yune Leong Futh Church 5. Choonghee Dhong Thien Houe Church 6.Nam Soon Church 7. Ling Liang Church 8. Chon Nee Than Alms House.
SOCIAL CLUBS: 9.Sea Ip Club 10.Toong On Club 11. Gee Hing Club 12.Sea Voi Club and Hupeh Association 13.Nam Soon Club 14.Han Fook Koon 15.Golden Dragon Club 16. Hoong Fat Koon 17.Low Fun Koon. 18. Chun Chun Thong 19. Ho Hei Koon.
EDUCATIONAL: 20. Chien Kuo School 21. Ling Liang School
COMMERCIAL: *I. Shoe shops* 1. Shoe shops 22. Ahon Bros 23. Wam Tong Shoes 24. Indo China Shoes
II. Restaurants 25. Pou Hing 26. Hao Ming 27.D'ley 28.Tung Nam 29. Chinese Breakfast street
III. Beauty Parlour 30. Lily Beauty Parlour 31. Karishma Beauty Parlour
IV. Saune shop 32.Pou Chong Brothers 33. Sing Cheung 34. Sing Ho
V. Wood workshop 35. Chen Carpentry & 3 associated workshops 36. Chen's Furniture-makers & Interior Decorators 37. M/s The Unique Carpentry
VI. General 38. Hap Hing Co. Chinese Provision & Medicine Store 39. Noodle Making units (2 nos.) *VII. Dentist* 40. Dr Mao Chi Wee Dental Clinic

Figure 11.2 Architectural Resource Survey of Chinatown indicating the diminishing holdings of the community and the transformations in the neighbourhood.
Source: Kamalika Bose, CEPT Summer School 2014

The history of the Chinese in Kolkata spans across four broad categories and landmark events. Starting in the 18th century, the first wave was marked by the setting up of sugar mills in 1778, by the first 'Cheeni' (person of Chinese origin; also Bengali for 'sugar'), namely Yang Dazhao, now endearingly called 'Atchew.' Today, Achipur – a fringe town near the Kolkata harbour, has become a pilgrimage site for the community. The second wave of settlers started arriving 1830s onwards, their migration triggered by successive famines and frequent uprisings against the corrupt Manchurian Qing dynastic rule. Overlapping with migration to California during the gold rush, Kolkata was an attractive port city en route to Beira in East Africa and further to Europe. In the third phase, the Chinese population peaked in Kolkata at 26,250 during the post-WWII period as a direct fallout of civil wars in China, and the Japanese invasions in Southeast Asia (Xing and Sen 2013). Anchored within central Kolkata's Tiretti Bazaar, Chinatown grew as the first formal settlement in South Asia.[2] The enterprising community, comprising Cantonese carpenters and boat builders from Guangdong, Hakka leather tanners, and shoemakers from Meixian County, dentists from Hubei province, and Shandong silk traders, carved their niche market to thrive in Kolkata's economic heyday.

The 1962 Sino–India border conflict, as the fourth key marker, rapidly deteriorated bilateral ties between India and China and irreversibly altered the community's future.[3] Unsuspecting Indian families of Chinese origin faced a backlash akin to Japanese-Americans in the United States after the bombing of Pearl Harbor. Some 3,000 Indians of Chinese descent were imprisoned and interned at Deoli Detention Camp in Rajasthan from 1962–1966, several framed as spies or Communist sympathisers (Ellias 2015). Passport seizure, cancellation of trade licenses, and confiscation of property were some further fallouts, along with the social stigma and mistrust that underlined their later release and rehabilitation. The disillusionment stemming from this victimization, discrimination, and loss of national identity and equal opportunity propelled a wave of mass emigration to Canada, the UK, and Australia in the 1970s (Ma and D'Souza 2020). Those that decided to remain in India, resettled in Chinatown and suburban Tangra to lead quiet, inconspicuous lives while rebuilding their compromised identity. Internment and migration have had other physical impacts within Chinatown over the last five decades. Newer communities – predominantly low-income Muslims and Bihari migrants – have moved into the vacant or sealed properties as multi-storey tenements seen along Damzen Lane and portions of Chattawala Gullee. Land seized by the government was developed into large office towers such as the Kolkata Improvement Trust and Poddar Court on Lhu Sun Sarani.

The most pressing question today is about stakeholders interested in the preservation of Chinatown's ethnic heritage. Additionally, what ends will restoration of these isolated historic buildings and structures serve without a long-term maintenance and reuse plan that is integral to local communities? The State Government of West Bengal does not allocate any budget for restoration of private properties even if they are listed on the 'Graded List of Heritage Buildings 2009' by the municipal corporation. They have to be privately funded or supported by central schemes or through international efforts. This calls into question the future of a significant quantum of Kolkata's historic buildings linked to ethnic populations. The Cha Project (Cities • Heritage • Architecture) was initiated to address these challenges, using a community-centric urban strategy that synchronises Chinatown's closely held formal and associational values with an integrated, forward-looking toolkit for its contemporary relevance and continuity. The project's appropriate implementation and outcomes could offer lessons for urban preservation in Kolkata and serve as a model for strengthening diversity of cultural experiences within Indian cities.[4]

From perception to preservation challenges: Chinatown within the mosaic of Kolkata's historic neighbourhoods

Before delving into project methodologies, it is helpful to broadly understand preservation issues in Kolkata today to contextualise the Cha Project's core objectives and processes. To the emigrant, Kolkata is a 'portable city of the mind' that is beyond the reach of urban blight and civic neglect and flourishes all the more as the real one declines (Suraiya 2009). To the resident it is a shadow of its erstwhile self, as the former capital of the British Empire in India[5], once counted as the grandest, most advanced of Asian cities. The city and its built heritage are today caught in this perceptual dichotomy, between a nostalgic self-image and a grim urban reality. Envisioned as the 'City of Palaces' in the late 18th century, the sobriquet lends a scale and monumentality to the idea of Kolkata that tramples its 'everyday urbanism,' one where architecture forms the backdrop of life and people, and is not the primary spectacle (Mehrotra 2001). This attitude, despite slow change, has also severely impacted the city's approach to the preservation of its historic neighbourhoods, including ethnic precincts such as Chinatown. Much of the effort continues to spotlight the former colonial precincts making preservation practice exclusive, inequitable, and entrenched in prejudices which are strongly dismissive of 'native' and 'ethnic' urban histories.

For Kolkata, the West Bengal Heritage Commission Act of 2001 and Section 2 (42A) of the Kolkata Municipal Corporation Act 1980 are two key pieces of legislation pertaining to the city's built heritage. Provisions under the Acts are carried out, in varying capacities, by three administrative bodies: the Kolkata Municipal Corporation (KMC), the Kolkata Metropolitan Development Authority (KMDA), and the Department of Municipal Affairs of the Government of West Bengal (DMA). The KMC's Heritage Conservation Committee grades and lists heritage buildings and recommends their conservation. There is a separate fund to ensure maintenance of

Grade I heritage buildings (currently 611 of 917 listed buildings are Grade I properties). Apart from individual grading of heritage buildings, as of June 2020 there are no existing systems that allow designation of historic precincts or conservation districts (Bose, 2013).

Most city neighbourhoods are additionally plagued by outmoded physical infrastructure and inadequate building safety codes, which present a hurdle to any preservation initiative. Suffering decline and neglect with overcrowding in tenement housing, building collapses, freak fires, waterlogging and poor drainage, such conditions have in turn lowered property values and diminished interest for new investment in the area. Where the State is ambivalent, it has often been local non-profits, preservationists, and an active civil society that have built awareness campaigns and pressured public agencies into mobilising resources that prioritise preservation. In Kolkata, sustained civil society action and agency has been absent until very recently. Be it the intelligentsia or the masses, the city's famed culture of protest has rarely voiced a strong resistance on heritage matters such as poor preservation practices, lack of government initiative, and systematic loss of historic fabric.[6]

The absence of economic returns from rehabilitation has made it an unattractive proposition for property owners and investors alike. Market-based incentives offer a powerful reason for the public at large to retain, care for, invest in, and responsibly rehabilitate historic buildings. With historic preservation largely perceived as being overly regulatory and authoritarian, financial incentives offer a positive and more broadly defensible reason for the public to get involved with responsible preservation activities (McLeary 2005, 1). With state agencies not prioritising incentives for repairs, restoration, and reuse, private initiatives alone have been ineffective in maintaining the abundant historic stock. This has resulted in a large inventory of unprotected, over-regulated private holdings. Restoration of individual buildings occur through isolated initiatives while the landscapes they are set in continue to fall apart (Mehrotra 2001, 161). Obsolete rent control and tenancy legislations further make maintenance and repair unattractive for landlords, often turning to the real estate sector to dispose of or suitably 'redevelop' their 'heritage' liabilities.

These factors underscore the urgency of community-led preservation campaigns that are neighbourhood-centric and formulate bottom-up approaches in converting challenges into opportunities for systemic change. A foundational shift in preserving Kolkata's architectural inheritance and legacy stems from understanding the nature of the native and ethnic enclaves – which were not a result of royal, religious, or colonial patronage. Entirely created by a composite mix of the migrant trading class, native professionals, and the mercantile elite, they need to transcend the established approach of 'landmarking' and merit new systems that celebrate their continuity into the future. The Cha Project therefore attempts to offer a model framework for undesignated (and unprotected) historic neighbourhoods in the absence of codified heritage regulations protecting them.

(Not just) architecture: Chinatown and the Cha Project's preservation paradigm

Kolkata Chinatown's architecture is strongly interwoven with the social order and migrant history of the community. To an expectant visitor, its discreet disposition and seemingly unremarkable buildings may come as a disappointment. Yet with a nuanced lens and informed understanding of migration patterns and resettlement forces in colonial India, the distinctions become self-evident. As a migrant minority's claim on a city through the built fabric, Chinatown systematically shies away from exuding traditional urban form including decorative elements associated with the parent culture, especially on exterior facades. Instead, the adoption of neoclassical facade-making principles and elements that adhere to the political climate and therefore architectural choice of the colonial administration is a guiding factor. Symmetrically placed Ionic columns on the Toong On Church facade and the arched entryway with Corinthian columns at Choonghee Dhong Thien Haue Church are representative of this phenomenon (Bose and Hilberth 2014). Yet it is the interior space of each temple that strongly expresses Chinese culture and characteristics through spatial elements, use of colour, calligraphy, artefacts, and festoons, especially in the shrine room. The ornate altars of Sea Ip and Sea Voi churches exemplify these qualities. Furthermore, the post-1962 detention phase gave rise to an architecture of camouflage – discarding established frameworks of single-use building classifications to create self-contained, multi-use typologies. The 17 Tiretta Bazaar Street, a monolithic structure with bare walls, arched windows, and a neoclassical entrance is a telling example of this pattern. The building houses the Choonghee Dhong Church on the upper floor, two noodle-making units, a warehouse, and two residential quarters on the lower floor: all of Chinese occupancy and ownership. Continuity in uses thereby becomes a tool for asserting and safeguarding ethnic identity over stylistic and visual expressions of culture (Bose 2016).

In seeking to revitalise Chinatown, therefore, the first unlearning for architects and preservationists – and a difficult one at that – is the acknowledgement that historic architecture is not the sole spectacle. The built form is just one strand of a complex tapestry of socio-cultural parameters that together represent the true ethos of the neighbourhood. Saving the built fabric, while a necessary first step, is alone insufficient in holistically reviving Chinatown. For that reason, the Cha Project adopted the Historic Urban Landscape[7] approach in accordance with UNESCO recognising that tangible built heritage and artefacts provide a strong context to the intangible heritage of the community. Life and activities on the streets are firm indicators of a traditional culture still alive that has intermeshed with narratives of change to shape their contemporary identity. Storytelling, performing arts, social practices, rituals and festive events, knowledge systems and practices, and traditional skills form a key part of this identity. Their cultural

Figure 11.3 The sacred and secular worlds seamlessly intersect to create multi-use spaces that form the social backbone for the community. Seen here is the Geehing temple and social club with daily morning mahjong games underway. Source: Thomas Hilberth, CEPT Summer School 2014

continuity without being reduced to tokenism or exoticism is critical for integration with contemporary Kolkata (Figure 11.3).

Business development and self-sustenance were the next steps as economic motives and enterprise have shaped the experience of Chinese in the city who have been identified with proficiency in a number of native trades. Today the numbers of independent family-run businesses have dropped but the commercial activities in Chinatown continue to retain, in its essence, the diversity and skill of previous generations (Bose 2016, 40). All traditional trades, either as retail outlets and workshops, remain present within a 1.5 km boundary of Chinatown signifying continuity in knowledge systems. This offers opportunities to build new strategies to scale local trades and the neighbourhood economy and incubate new businesses. Simultaneously architectural restoration of the six Chinese temples, listed as Grade I heritage and the most visible and compelling physical assets, would be prioritised. The sparse remains of secular architecture play an equally important role in representing identity and living practices. In their absence, the temples, as isolated monuments, become mere relics in an altered urban landscape, bereft of context and meaning. Restoration and repurposing these with viable economic uses become core agendas for Chinese built heritage while sympathetic new design interventions on vacant and under-used plots would introduce and improve essential urban amenities that are currently lacking.

Inclusion of non-Chinese communities and new residents in a diverse demographic and mixed land-use is crucial for a holistic approach. Keeping the Chinese community at its heart, the process engages with stakeholders of public and private sector companies occupying modern high-rise buildings at one end, and street encroachments by the urban poor at the other. Under the current circumstances, the old and new must establish a harmonious coexistence in an effort to enhance the historic value of the area – for residents, workers and visitors alike, through an integrated approach to urban conservation and community planning. Preservation objectives are ineffective when they disengage with broader issues of urban planning, infrastructure upgrade, participatory processes and social justice in historic neighbourhoods. Integrated planning that addresses traffic management and pedestrian safety, sanitation and drainage, encroachment and hygiene, along with contextual urban design, form the parallel pillars of the Cha Project's revitalisation paradigm.

Testing the paradigm: strategies and processes towards implementation

With a mandate of bringing together scholars, historians, architects, designers, preservationists, urban planners, artists, private enterprises, government bodies, the community, and NGOs to galvanise a community-driven, heritage-oriented revitalisation of this last surviving bastion of the city's multicultural past, the Cha Project evolved a process model that is under implementation in various phases (Detailed Project Report 2014, 7). A three-pronged integrated approach of urban revitalisation, environmental improvement and economic empowerment has informed the key pathways. Some of the core strategies and processes are enumerated below:

Architectural resource survey and recommendations

A survey and documentation of Chinatown's urban heritage and architectural inventory became the first systematic method to identify, document, and evaluate the neighbourhood's cultural resources. A critical section of documentation was conducted as part of CEPT Summer School 2014, a collaborative workshop between CEPT University in Ahmedabad and Aarhus School of Architecture, Denmark, led by the author. The joint exercise produced a publication titled *Cheenapara: Cultural Identity & Urban Heritage of the Chinese in Calcutta* for wider circulation and awareness building within the community and externally. Additional documentation, entailing streetscape and individual property surveys to establish architectural integrity and opportunities for reuse were subsequently undertaken. A set of recommendations and guidelines were evolved and compiled as a project report

(Detailed Project Report 2014), which was submitted to the Government of West Bengal in August 2014.

Streetscape restoration and adaptive reuse

The land use survey further identified a set of buildings within the community, which are inaccessible, misused, underused, or vacant. Programming that unlocked the economic potential of such buildings was explored as a tool to enhance the social, recreational, and tourism potential while introducing self-sustaining revenue models. With minimal new construction undertaken, plans were drawn up for facade restorations of existing assets along Sun Yat Sen Street, to reinstall pride and identity for residents, restore businesses in Chattawala Gullee, and be an attractive point of entry for tourists. As an accretive evolution of diverse Chinese sub-groups who came together, over centuries of hard work, trust, and social enterprise to contribute to Kolkata society and its cosmopolitanism, the streetscapes remain humane and without hubris. The preservation approach employed therefore rejects pastiche and cosmetic treatments that transplant global and commercial notions of touristy Chinatowns. Blackburn Lane and Chattawala Gullee were identified as action areas to implement this strategy. Both lanes have retained, in ownership and typology, the largest concentration of buildings that directly relate to old Chinatown's history and contemporary narrative. It contains religious, social, cultural, and commercial typologies within this narrow street – and showcases the diverse cross-section of spaces integral to community life. In an amorphous urban fabric that characterises *Cheenapara* today, Blackburn Lane therefore attains high value and significance to the revitalisation project and its future. The untapped potential of historic structures within Blackburn Lane and Chattawala Gullee, e.g., Hupeh Association, Voi Ling club storage area (former dormitory), and the Chonee Thung Alms house are being leveraged through reprogramming for suitable new uses such as interpretation centres, home-style restaurants, community-run retail units aligning with the economic revival plan (Figure 11.4).

Revival of the community-driven micro-economy

Facilitating economic development for the community through livelihood opportunities is a key pathway for boosting the micro-economy. Often the absence of economic engines that drive the revitalisation process become core deterrents in steering a value-for-all approach. The Cha Project addressed this by first enhancing the local economy by boosting existing native trades and traditional skills while simultaneously incubating allied new ventures and entrepreneurship. An organised food street and night market along Chattawalla Gullee, host of the morning breakfast, is part of the first phase of economic development involving younger members in its administration

Figure 11.4 Before and after rendering of a streetscape corner at Chattawalla Gullee, the site for morning Chinese breakfast and street vendors. Source: The Cha Project, Detailed Project Report 2014, 129

and management. The second phase of this initiative hinges on astute market research that identifies and creates a new network for existing and potential enterprises through business development strategies. Diversification of native skill sets such as shoe-making and carpentry that allow access to newer markets, survive competition, and address the new aspirations of community youth is the core objective. Creating alternative livelihoods is envisioned through promoting architectural and cultural assets of the community. Enhancing visitor experience through heritage trails, creation of an interpretation centre, and giving visibility to local crafts are key tools to create ownership and income. Walking tours and interpretation facilities led by women and youth from the community, offer alternate sources of livelihood and dovetail into the efforts at creating local businesses. Visitor engagement thus helps boost the micro-economy, revenue generation from which can be ploughed back into preserving the physical fabric of Chinatown.

An inclusive model using social capital credits

As discussed earlier, the urban boundaries of old Chinatown integrate and absorb multiple ethnicities with economic disparities, who share space and livelihoods alongside the Chinese community. It is necessary to be inclusive of such alternate socio-cultural voices as well to avoid insular narratives of a singular community. Revitalisation cannot be envisaged without including the other communities and the marginalised population. The presence of a strong parallel informal economy coupled with inequities stemming from destitution, homelessness, and squatting have created a social imbalance. The Cha Project does not view eviction as a way forward and hence has partnered with Asia Initiatives in New York to implement SoCCs[8] or Social Capital Credits, a transformative system of exchange that could potentially minimise poverty on the streets of Chinatown. Residents can earn SoCC points by participating in community activities or helping in community initiatives, like collecting garbage, painting public spaces, or building a green wall. These credits can then be used for goods and services within the community, such as basic household items at the local market or to subsidise education.

Consolidating urban amenities: waste management

Improper solid waste management has emerged as the bane of modern India today. Despite national level campaigns, policies, and incentives to minimise the menace and with local municipalities attempting to remedy earlier ills, it has been a slow and cumbersome path to progress. Chinatown's historic resources are not only threatened, but are at serious risk owing to the presence of both garbage disposal vats and a municipal waste collection centre located in close proximity to several historic buildings and Chinese temples. Advocating for the removal of such sites while installing a garbage compactor unit and implementing a sustainable closed-loop waste management plan for Chinatown forms a part of the broader project plan.

The Cha Project: lessons for practice through its challenges and successes

Chinatown's urban heritage today thrives as a multicultural, multi-ethnic, and richly textured ensemble whose true potential as a cultural and economic asset to Kolkata and its people has long been unappreciated and unexplored. The Cha Project recognised these aspects and attempted to introduce innovative, yet pragmatic strategies in urban revitalisation that are contextual and deeply embedded in local community values. Such holistic initiatives are aimed to surpass cosmetic beautification of the built heritage to become strong vectors of change by strengthening the local economy, fostering social ties and making the city more liveable for locals and attractive for visitors. And yet there have been consistent challenges that have impeded the pace of implementation. Apart from formulating strategies and pathways discussed earlier, those that enable a socio-economically driven urban conservation paradigm, the Cha Project's ongoing challenges and small successes also offer lessons for practice in diverse contexts and communities. These lie in the areas of workability of the public-private partnership (PPP) model in unlisted and unprotected heritage precincts, funding toolkits and their translation into phased cash flows for the project, and community mobilization from the bottom up.

Despite the project report being officially submitted to the Department of Tourism, Government of West Bengal and forwarded to the chief minister's office in 2014, to enable action and implementation of PPP projects, progress on-site has been slower than anticipated. Two major hurdles thus far have been impediments in securing and mobilisation of adequate project funding. The project plan identified five core funding strategies over the first three years – corporate social responsibility, institutional/foundational grants, state government funds, private donations and philanthropy, corporate sponsorship, and event/merchandise-based fundraising. Funds raised are to be managed by the Cha Project under a management partnership as an integrated, area-based, multi-stakeholder approach. A decentralised mode of working between Singapore and Kolkata and the lack of a dedicated project team with targeted fundraising goals and campaigns has had a negative impact, especially in the absence of a singular patron through all phases.

Second, low institutional will and political inertia has significantly slowed down implementation and pushed back timelines. As a relatively small population still remaining in India, Indians of Chinese descent have become minorities even within the broader definition of minorities. Forming an insignificant 'vote bank' (group of loyal voters), minimal presence in minority forums, and no state legislative assembly representation, they thrust minimal or no pressure on the broader political landscape of West Bengal. While gestures of beautification and urban upkeep are undertaken by municipal agencies and state authorities periodically, there is a huge political inertia at addressing the larger concerns of cultural continuity. The Cha Project's integrated approach entails a significant component of urban amenities and infrastructure improvements which also require the city agencies to be in the driver's seat. There have also been instances of disconnect between the project report's recommendations (Detailed Project Report 2014) and on-site interventions by the municipal corporation where they have been made. For instance, the location, incorrect placement, and poor management of a garbage compactor to address the solid waste management issues in the neighbourhood have compounded problems for the Toong On Church, a Grade 1 heritage building. Spillage and stench from the compactor has made the surrounding context and street to Toong On unsightly and difficult to approach, thereby negatively impacting visitor inflow.

And yet the collective energy and spirit of Indians of Chinese descent has gone from strength to strength since the project idea was germinated. Undaunted by obstacles along the way, unprecedented community cohesion is paving the path for self-initiated, community-led efforts that today transcends factionalism and clan tussles. A stronger and united leadership under the Indian Chinese Association for Culture, Welfare, and Development[9] along with the temple associations, are actively lobbying with state authorities, improving consular and diplomatic ties, leveraging media attention and seeking a reach and visibility for Chinatown that is beyond its boundaries. The Cha Project has opened new frontiers for self-expression with a surge in community pride and confidence to enable the forging of new cultural collaborations. The recently launched Community Art Project, annual Dragon Boat Festival, and the *Cheenapara* Heritage Festival and Night Market are testimony to that. In January 2018, members of the Seu Vui Association and Voi Ling Club self-funded the adaptive reuse of a former dormitory within their club premises into a family-run restaurant in Blackburn Lane. Through these community-driven initiatives, Indians of Chinese descent are, slowly but surely, regaining their voice and carving a renewed presence for their towering legacy in the city they call home. Despite having lost crucial time, it is yet to be seen how the Cha Project achieves its core objectives – either partially or in full. The community is adequately enthused and it is imperative that tangible action be witnessed on-ground for energy and optimism to not wane. The seeds of systemic change have been sown and the Cha Project's success can be a watershed moment for urban conservation and inclusive development in the region.

Notes

1 The 19th century temples, consecrated to warrior gods and Lord Buddha have a nomenclature of 'Church,' as a remnant of colonial city surveys. Ethnic identity was concealed to attain greater legitimacy and perhaps avoid desecration.
2 In the 1920s a second Chinatown emerged in the eastern suburbs of Tangra where the Hakka Chinese who specialised in leather trade set up a flourishing business of tanneries. Chinatown at Tiretti Bazar remained predominantly Cantonese and this distinction continues until today.
3 Struggles faced by the community as an aftermath of the Sino-India War has been deftly portrayed by documentary film-maker Rafeeq Ellias in the award-winning "The Legend of Fat Mama" (E, 2010) and "Beyond Barbed Wires, A Distant Dream" (2015).
4 *Chiya. Chaha. Chaya. Theneer, Tī.* Different words for tea in different parts of India. Interestingly, in Bengali it is Cha as it is in Chinese. Although CHA primarily stands for Cities, Heritage and Architecture, it takes on an acronym

that defines the essence of the project – a revival through teashops and cafes. Tea has a rich legacy of bringing people together, which served as a metaphor for the Cha Project's spirit. Initiated by a Singapore-based consortium, the CHA Project works in association with the Indian National Trust for Art and Cultural Heritage (INTACH), Kolkata Regional Chapter.

5 Kolkata (formerly known as Calcutta), was founded in 1690 as a port city in Eastern India and British trading post. It served as a capital of the British Empire in India from 1757 to 1911, after which the capital relocated to New Delhi (Bose, 2013).

6 Calcutta Architectural Legacies (CAL) is a citizen's initiative spearheaded by author and musician Amit Chaudhuri in 2015, with the aim of preventing the destruction of the city's built heritage, particularly from the late 19th century and the first half of the 20th century. It also advocates for the designation of historic precincts.

7 As stated by UNESCO: The historic urban landscape approach moves beyond the preservation of the physical environment, and focuses on the entire human environment with all of its tangible and intangible qualities. (https://whc.unesco.org/en/news/1026/)

8 Social Capital Credits is an alternative system of exchange which uses social instead of financial capital. The scheme allows residents to trade actions and processes which have a positive social benefit for goods and services not accessible to them through traditional financial economy (http://asiainitiatives.org/urbanrural-initiatives/cha-project-kolkata-india/).

9 Indian Chinese Association for Culture, Welfare and Development (ICACWD) is registered as a society under the West Bengal Societies Registration Act of 1961 and approved by the Government of West Bengal on 4/10/1999.

References

Bose, Kamalika. 2013. *Incentivizing Urban Conservation in Kolkata: The Role of Participation, Economics and Regulation in Planning for Historic Neighborhoods in Indian Cities*. Ithaca: Cornell University, Master's Thesis.

Bose, Kamalika. 2016. *Invisible Identities, Uncertain Futures: Upholding the Cultural Heritage of Kolkata's Chinatown*. Marg Publications, March, Volume 67, Number 3.

Bose, Kamalika, and Thomas Hilberth. 2014. *Cheenapara: Cultural Identity & Urban Heritage of the Chinese in Calcutta*. Ahmedabad: CEPT University.

Ellias, Rafeeq. 2010. *The Legend of Fat Mama*. BBC documentary.

Ellias, Rafeeq. 2015. *Beyond Barbed Wires, A Distant Dream*. Self-produced documentary.

Ma, Joy, and Dilip D'Souza. 2020. The Deoliwallahs: The True Story of the 1962 *Chinese-Indian Internment*. New Delhi: Macmillan.

McLeary, Rebecca. 2005. *Financial Incentives For Historic Preservation: An International View*. Philadelphia: University of Pennsylvania, Master's Thesis.

Mehrotra, Rahul. 2001. "Bazaars in Victorian Arcades: Conserving Bombay's Colonial Heritage", in *Historic Cities and Sacred Sites – Cultural Roots for Urban Futures*, edited by Ismail Serageldin, Ephim Shluger, and Joan Martin-Brown. Washington, DC: World Bank, 154–163.

Nagpal, Swati. 2012. "The Gomti Riverfront in Lucknow, India: Creating a Landscape of Heritage", in *Heritage And Development:* Papers Presented at the

12th International Conference of *National Trusts in December 2007*, edited by Asha Rani Mathur. New Delhi: INTACH.

The Cha Project: Detailed Project Report of a Heritage-led Revival of Kolkata Chinatown, Submitted to Department of Tourism. Government of West Bengal, November 2014.

Suraiya, Jug. 2009. "Calcutta Chromosome", *Outlook Magazine*. http://www.outl ookindia.com/article.aspx?239452, Accessed March 3, 2017.

Xing, Zhang, and Tansen Sen. 2013. "The Chinese in South Asia", in *Routledge Handbook of the Chinese Diaspora*, edited by Tan Chee-Beng. New York: Routledge, 205–226.

12 Saving the history of Malabar mosques and their communities

Patricia Tusa Fels

Introduction

The Malabar Coast, now part of the Indian state of Kerala, has captured the imagination of travellers and geographers for centuries. Cut off from the rest of India by the mountainous Western Ghats, the region has a unique history. Malabar was famous throughout the medieval world (and even Roman and Greek chronicles) for its spices[1] and open trading ports. People of the coast were little influenced by the "official" narrative trajectory of India. No invasions came from far away, no world empires rose and fell, until the arrival of Vasco da Gama in 1498 brought Western colonialism[2]. The very identity of the unique Malabar culture is the melding of influences from the West (the Mediterranean), Arabia, the East (China and beyond), and India. Trade in Malabar was instrumental in the "...communication of ideas, cultures, technologies, and religions, the 'invisible cargoes' delivered and consumed along with commodities" (Riello and Roy 2009, 8).

Unlike much of the rest of predominantly Hindu (79.8%) India[3], Kerala does not have a dominant religion. Christians, Muslims, and Jains come close to equalling the number of Hindus. Christianity and Judaism established outposts on the Malabar Coast in the 1st century CE. Some of the first Muslim settlements in India occurred in the Kerala ports of the 7th century. The open environment of Kerala nurtured accommodating forms of Christianity[4], Islam, and Hinduism. Adherents of Christianity, Judaism, and Islam modified local Hindu customs to accommodate new converts.[5]

The vernacular architecture of Kerala reflects the geography, climate, and resources of Malabar – and the history of the interwoven cultures. With two annual monsoons, the mountainous rainforest produces an abundance of tropical timber. As traders arrived and settled, they adapted the indigenous building vocabulary to new uses. This chapter will examine the built heritage of the Muslim traders, particularly the mosques built along the Malabar Coast. These mosques took advantage of local materials and met religious liturgical necessities utilizing the building traditions of Kerala (Figure 12.1).

Figure 12.1 View of Miskal Mosque in 2015. Source: Author.

Up to the early 20th century, mosques in Kerala were built in the style of the local vernacular and faithfully maintained. The prominent elements of the Kerala building tradition were the ubiquitous use of tropical woods and the dominance of large elaborately carved roof structures. The force of the monsoons and the need to shed large volumes of water led to grandly scaled pitched roofs, crafted from the woods of the nearby forests. The pyramidal roofs of palaces and temples were not just re-imagined for mosques, but also for homes and warehouses. Large expansive embracing roofs implied a prestigious owner. Timber was used for columns, roof structure, beams, eaves, brackets, ventilation screens, ceilings, and more. Most elements were carved and intricately joined. Centuries of boat and building construction meant that there was a skilled population of woodworking artisans. Since the monsoons brought heavy rains and frequent flooding, builders elevated structures, typically on a granite base. Walls were open or built of the local, porous, and easily worked laterite. The massing of buildings reflected the belief that no structure should exceed the height of the ubiquitous coconut palm. All structures displayed ingenious means for cooling. Carved roof gables, louvred upper floor walls, interior courtyards, and exterior verandas are all elements of the Kerala vernacular.

Despite the rich layered urbanity of Kerala's port cities, no city, regional or state government has been able to implement any real enforceable regulations to protect the historic urban fabric. Often it is only the whims of the marketplace that have conserved heritage by default. Several European (colonial) structures in Kochi have undergone rehabilitation, transforming them into boutique hotels, eateries, and stores. Even with file cabinets full of reports by UNESCO, INTACH, the Dutch government, and others, there exists only a sketchy inventory of historic buildings.[6] None of the local vernacular houses or the historic godowns (warehouses) are even mentioned in the voluminous reports.

There has been little investigation of the ensemble of the port communities: streets, waterways, homes, shops, religious buildings, and warehouses. The focus of historic preservation efforts has been on protecting individual buildings as monuments. UNESCO reports in 2009 and 2011 list three mosques in the historic area, but none are protected or even proposed as such (UNESCO New Delhi Office 2009). The urban centres of Kerala port cities exhibit a richly textured environment where the significance lies in the whole more than in individual pieces.

Although Kochi's oldest mosque (Chembittapalli) and the Paradesi Synagogue both date from the 16th century, the UNESCO report concentrates only on the synagogue and Jew Town (UNESCO New Delhi Office 2009). There is little recognition of the surrounding older Hindu and Muslim centres, which remain large and vibrant. The synagogue in Kochi, a marker of an important history, has been highlighted for conservation. Unfortunately, there is no longer a Jewish community left in the area. The messy world of religion and real people also tells vital stories. Historic buildings that are presently unprotected are essential in providing physical evidence of the city's history. The legacy of the Malabar trading cities is steadily disappearing. A victim of insensitive modernization or demolition, this "living" heritage in all its diversity defines the history and culture of Kerala.

> Protecting the unprotected architectural heritage and sites ensures the survival of the city's sense of place and its very character in a globalizing environment. It offers the opportunity not only to conserve the past, but also to define the future.
>
> (Centre for Heritage, Environment and Development 2013, 11)

Mosques and community settlements

> Arabic-speaking Muslims.... were also found by the 8th century as traders along the Malabar coast of the southwest, where they settled, intermarried, and sustained distinctive cultural forms forged from their Arab ties and local setting, and in so doing helped link 'al-Hind' to seaborne trade routes.
>
> (Metcalf 2002, 5)

Two of the most interesting historic Muslim communities on the Malabar Coast are Kuttichira, in Kozhikode (old Calicut), and Kochangadi, in Kochi (the old Cochin peninsula). Both are settlements born from international trading. Both share common areas with adjacent Hindu and Christian neighbourhoods. In both, 500-year-old mosques continue to function. The ensembles of mosques and their surroundings are a unique and irreplaceable resource, not only of architectural heritage but also as proof of centuries of co-existence. In the 1500s, the Muslim communities had great wealth, as European colonialism had not yet eaten away at their trade. Religious and community leaders joined to build exceptional structures that showcased the artisans of Kerala, using the abundant woods of the adjacent tropical forests, and backed by the wealth of the trading communities (Fels 2011, 29). Examining these two communities today can tell us much about the role and the practice (or non-practice) of conservation in Kerala. The condition of the traditional housing and the conservation of the historic mosques tell the story of how distant, even irrelevant, conservation is for the daily life of Kerala.

The Malabar Coast is a long way from Delhi and the headquarters of the guardians of Indian historic preservation: the Archaeological Survey of India (ASI) and the Indian National Trust for Art and Cultural Heritage (INTACH). Both the ASI and INTACH have been focused on an archaeological dig at Muziris[7], the ancient port of the Arabian Sea that was washed away in the 14th century to be replaced by Kochi. This excavation is of great consequence and has unearthed an enormous amount of data. It was continuous from 2007 until 2015 when work was suspended because of complaints by far-right religious extremists.[8] Not just at Muziris, but throughout India and Kerala, the volatility of religion has caused the loss of important heritage. If the archaeologists who dig have difficulties, those who want to protect the steadily disappearing built fabric above ground face an even bigger challenge. The living heritage present in the historic settlements of Kochi and Kozhikode can be judged to be of the highest calibre for architectural character and cultural importance, yet it has little to no protection.

A generally conflicted view of the old city has led to a tendency to build anew. The local communities have an ambivalent attitude towards historic structures. In the recent past old mosques have been demolished, or updated with little or no awareness of their historic importance. The general population widely believes that it is impossible to update old structures with modern amenities. In addition, there is a long-standing belief of the poverty of the Malabar Muslim neighbourhoods, which perhaps translates into a belief in the poverty of the architecture. On the contrary, these everyday neighbourhoods reflect a rich built form of homes, shops, schools, water tanks/ pools, and mosques built with artisanal care and pride.[9]

For over 500 years the old mosques of timber and stone endured, fighting off bugs, mould, and rain. Most were still standing until recently, even as ongoing upkeep was no longer undertaken with any regularity. Venerable old mosques are endangered by abandonment, outright demolition, or remodels that wrap the original structure in concrete and kitsch. The uniqueness of the Malabar mosques, instead of being celebrated, has become suspect and hidden for its connection to the local culture and customs. People will speak lovingly of their old mosque, the coolness of the hall, the richness of the wood, but they seldom reject calls for a new mosque of concrete and conditioned air.

Why is there a lack of interest by "official" government conservation groups? Why does local government fear involvement with religious communities? One reason for this fear and lack of interest can be seen at the Muziris dig. Though the site was not religious, fundamentalists saw the project as an attempt to belittle Brahmanical history. In fact, the only clear story that emerged from the excavations was Kerala's longstanding cultural syncretism. "The significance, symbolism and scope of the Muziris heritage project is far greater than simply reviving history because it highlights the very composite culture of tolerance and integration that gives not just Kerala, but all of India its place in the world" (Mirchandani 2015).

In addition, Kerala's long-standing connections to Arabia have far-reaching effects. Young men go to the Persian Gulf to work. Riyals and rupees return along with the aesthetic of Gulf modernism (and a less tolerant Islam). Glittering modern architecture with no connection to place is seen as trendy. Workers return to Kerala, tear down old family homes, replace them with "Gulf Mansions", and fund new mosques. The new flat-roofed concrete structures ignore the monsoons, the heat, and the local building traditions. Charles Correa speaks of the power of the mythic image of Islam coming from the desert: "For most people, these are the images of Islamic architecture that spring to the mind. Yet, ironically enough, the majority of Muslims do not live in that hot-dry belt from Algiers to Delhi, where this kind of built-form is prevalent. The vast majority ... live in hot, humid climates. What they need is not dense massing, but light, free-standing built-form, and cross ventilation" (Correa 2012, 65).

The mosques of the past with open verandas, high roofs, and naturally cooled interiors have been replaced with concrete domes, minarets, Disneyesque versions of the Arabian Nights and Mughal monuments. Peaceful interiors with the sea breeze blowing through the opened walls are disappearing; instead, air-conditioned marble palaces offer a cold hard surface for prayer.

Kochangadi (Kochi)

Kochi, a jewel of the trading culture, is the second largest city in the state (Kozhikode ranks third). Sited on a peninsula with the Arabian Sea to the west and a sheltered waterway to the east, the city is blessed with a lovely natural setting and protected port. Rich in history, the peninsula has remained somewhat distant from the intense development that has occurred on the mainland. Over the years the old colonial centre has become a tourist destination where historic residential and trading areas have been transformed into tourist destinations. Some 70% of Fort Cochin's houses have

Figure 12.2 The Bazar Road godowns extend to the water with a mix of courtyards, warehouse space and living quarters behind the street facing shops. Source: Author.

been converted to homestays or hotels in the last decade (The Hindu, May 24, 2015).

Mattancherry, directly south, and the adjacent Kochangadi, are the sites of the original trading communities. Here people from throughout the Indian Ocean set up homes and businesses. Bazar Road, with godowns along the waterside, is the link between Fort Cochin and Mattancherry. The marks of the spice trade are still visible; even today bags of spices arrive from the countryside and are sent to all corners of the world. But numerous godowns are being repurposed for tourism. Wholesale and retail businesses continue but owners feel threatened. While many of the godowns still stand as shown in Figure 12.2, several massive old structures have disappeared only to re-emerge as modern hotels.

The unique spatial relationship between what was the main trading area, the port and a series of churches, synagogue, temples, and mosques makes Bazar Road/Mattancherry an especially important historic enclave. The total ensemble is a walkable tour of local history through centuries. Generations of families lived here and were supported by trade; communities created temples, churches and mosques, schools and businesses.

The beauty of Bazar Road is one of complexity, displayed in a myriad of spaces. The funding to document the enclaves of various religious communities has been minimal, and local government has been unsuccessful in enacting regulations. The political will to create building conservation guidelines is missing. Most of the attention and the few regulations have been focused on colonial Fort Cochin and the blocks that contain Jew Town and the king's former palace.

There has not been a shortage of ideas for building reuse: a godown community centre for local youth, a training centre for jobs, godowns for offices and tech start-ups, and a museum sponsored by the Pepper and Spice Trade Association.[10] So far, ideas that highlight projects for the community have been shelved, and are often replaced by tourism projects. Divergent views exist: one group sees the area as a tourist destination, another views it as a past that is best forgotten, and another as a place of continuing business and home.

Moving south of Bazar Road one enters Kochangadi, the first and oldest Muslim settlement. The old housing of Kochangadi has mostly disappeared. Ten years ago, bungalow mansions with elaborate plasterwork and large timber rain-shedding roofs were still visible. Today, even as the lanes of Kochangadi endure, behind the walls new homes display none of the local vernacular. Carved wooden lintels, doors, and beams have been replaced with fake arches and metal doors. The salvaged carved wooden pieces are now for sale in galleries on Bazar Road.

In terms of urban spaces, two community tanks/pools remain in Kochangadi. One is behind Mammu Surka, a small older mosque that has survived intact. The other faces the jewel of Kochangadi: Chembittapalli – the copper-roofed mosque. Chembittapalli is known for its history, fine woodwork, striking wood columns – and of course the roof. In fact, it is considered to be "one of the most interesting surviving examples of Muslim architecture in South India" (UNESCO 2011, Appendix 3, p 10) (Figure 12.3a and b).

The mosque administrators have peculiar ideas of how to maintain their mosque. They have re-built stone walls with concrete and have placed two large additions in positions that overpower the old mosque. All views of the once regal carved entry façade have been obscured. Lime plaster has been replaced with cement plaster, which does not allow the laterite walls to breathe. Several young men of the mosque voiced objections to the renovation techniques but they were not heard. No conservation architect or city administrator would dare put forth an opinion. Instead of welcoming the assistance of professionals, the mosque administrators see it as an infringement on their religious freedom.

Besides these political social problems, there is a lack of knowledgeable artisans. The wooden columns, beams, and rafters of the mosque were put together with no nails, using mortise and tenon joints. Present-day carpenters have lost the ability to fashion nailless joinery for large timbers. The path that the leaders of Chembittapalli have followed is typical of many of the Malabar mosques. Disparate materials and disproportionate massing abound. Just down the road, the Calvathy Mosque was renovated with

Figure 12.3a and b Chembittapalli, 2013. Additions to the mosque have expanded from one story to two stories and cover the entire eastern façade of the mosque. Image above shows the original carved porch, now concealed by the additions. Source: Author.

domes and an addition that overpowers the original prayer hall. The so-called "first" mosque of India, in nearby Kodungallur, was clad in concrete, completely covering the original structure. At least these original mosques still remain albeit behind modern exteriors, versus the many that have been demolished. Another mosque in Kochangadi (Akathe Palli, along with the tomb of a cherished spiritual leader) was extant in 1990s but has since been completely rebuilt. The old shrine remains but its distinctive wooden gables are missing.

Kuttichira (Kozhikode)

Calicut/Kozhikode has not been at the centre of a development push like Kochi. There has been less pressure on the old areas of the city, but this is now changing and the first signs of loss of the historic fabric can be seen. Kuttichira is a particularly unique community. Three historic mosques (Mithqalpalli, Jami Masjid, and Muchandipalli) form a striking trio of vernacular Malabar mosques. Besides these impressive mosques, a central pond/tank, and a historic market area, Kuttichira holds a wealth of large family homes. These Tharavadu (*tarawad*), shown in Figure 12.4, are ancestral joint household homes where families live communally. Kozhikode is part of the zone of Northern Kerala that follows a matrilineal system. There are approximately 500 Tharavadu in Kuttichira. Typically, they are rambling two-storey structures with courtyards, wooden ceilings, and much decorative carving of wood components. Most of the houses have a front

Figure 12.4 The neighbourhood of family homes (Tharavadu) with walled courtyards is centred on the historic Miskal Mosque, seen at the upper right.

veranda with a *kottil* or raised platform for meeting visitors. A large compound is formed by walls, which at the street are marked by roofed entry portals.

All of these elements are typical of North Kerala. Both Nair Hindus and Mappila Muslims[11] follow the traditional matrilineal society. Houses pass from mothers to daughters. The groupings of Tharavadu present a powerful urban element. Lanes with high walls connect the compounds, mature trees abound. Air circulation in and around the building is insured by the walled outdoor space. House walls constructed of native laterite tend to breathe. The entire composition provides cool interior spaces. Residents note that the old houses are more comfortable than the new concrete houses.

Today, with joint families giving way to nuclear families, a desire for separate accommodation is on the rise. Interviews by architecture and planning students at the National Institute of Technology in Calicut found that 90% of interviewees wanted private homes with non-shared kitchens, 50% wanted to keep the old with changes, and 50% preferred demolition (Aneesh Manthanath 2015). In general, the women were more appreciative of the existing houses and wanted to keep them, albeit remodelled. In the last ten years, the number of concrete framed structures has greatly increased in Kuttichira. On some properties the old has been completely removed while others have boxy additions. According to a 2014 survey, 80% of Kuttichira structures had a sloped roof (Krishnakumar 2014). As the new structures rise there will be fewer gables and more flat tops.

There is no architectural or technical reason why old matriarchal homes could not be re-configured to provide a common space, along with four to six separate apartments having private kitchens and baths. There exist many straightforward ways to place plumbing in an old building. Throughout Europe, and beyond, old buildings are updated and transformed for contemporary living, complete with wi-fi, modern heating and cooling systems, high-design kitchens and bathrooms.[12] There is no need to demolish old structures in order to insert electrical wires and plumbing pipes. Stones are chiselled to provide pipe runs, and walls are re-plastered. Waterproofing and insulation can be added. At the end, there is a modern home, without the waste, environmental impact, and expense of demolition. In Kozhikode, few have ever seen an older structure retrofitted for contemporary living. They fear that water pipes cannot be installed in the walls, that electrical wires aren't allowed (even though most homes have already installed surface mounted electrical wiring), and that renovation will be more expensive than demolition and rebuilding. Kuttichira needs a demonstration project to convince the community that modernization is possible and affordable through renovation. Making that first step is extremely difficult. The national conservation experts work mostly on grander buildings and have little experience with remodels of domestic vernacular houses. The local builders prefer to demolish and rebuild.

Studies by INTACH Calicut on the Tharavadu are a helpful start (Meribha 2014). But Kozhikode has not received the level of recognition found in Kochi. The majority of documentation has been by well-intentioned design students who have conducted surveys of several houses and a series of interviews. Unfortunately, the option of re-purposing the common hall, repairing timbers, and re-arranging spaces to create separate apartments for the present inhabitants has not been explored. Part of the problem is the paucity of experienced conservation architects. Students do not have years of practice and many of their professors have not worked in private practice. Neglect of the 150-year-old houses is a serious problem, especially since many of the present-day residents are low income (the wealthier have moved away). Though most houses in Kuttichira have water, sewer, and electrical connections, broad upgrades have not proceeded over time.

The most prominent mosque of Kuttichira, Mithqalpalli (Miskal Mosque), centres the community, reflecting onto the grand pond ("chira" means tank, the typical large water body that is the focus of south Indian towns). Mehrdad Shokoohy designates the mosque "among the major monuments of Muslim India" (Shokoohy 2003, 153). Built by Miskal Nahod (or Nakhuda Mithqal), a famous 14th century merchant and ship owner, the mosque was standing in 1341 when Ibn Battuta, the traveller/chronicler extraordinaire visited Calicut. Miskal Mosque has undergone many rebuilding campaigns, but the form and the majesty of the many-tiered roof and the raised stone base all respect the original structure, a direct descendent of Kerala palace and temple vernacular.

The mosque underwent major maintenance in 2011–12. ONGC (Oil and Natural Gas Corporation) funded two projects: one at the nearby Thali Temple and the other at Miskal Mosque.[13] Both are historic properties and the grants insured that these buildings will continue to serve the community into the future. But grants such as these are extremely rare, especially by public companies that cannot fund religious activities. This one occurred despite the religious connections (as a public entity, ONGC had to receive a waiver to apply the funds). The projects happened because an influential patron/politician pushed them forward. The majority of the historic mosques of the Malabar Coast have no such enlightened benefactor.

Role of conservation

The first task for heritage leaders in Kerala should be to *educate* the residents of the neighbourhood on the economic and social values of conservation. Until the local communities see the value and importance of their structures, there is no hope of conservation. Overcoming political stasis is difficult and can only be successful with a sustained community effort behind it. To assist this task, the local city planners and heritage officials must respond to citizens' concerns and create guidelines and regulations with resident participation. A series of obtuse rules emanating out of Delhi or from further afield would only add to the present distrust of conservation. In Kerala, as in much of India, there is little awareness of the importance and uniqueness of local vernacular structures. With an effort by conservationists to bring the message to the people, local communities might stand up for their homes, mosques and temples. At present, such efforts are few. Building conservationists need a broader base, and should look to establishing more connections with environmental organizations. Climate issues cross boundaries and could align heritage and environmental goals. Promoting conservation at all levels (from buildings to forests), these two groups could highlight the principles of sustainability. Conservation maintains buildings that require little imported energy and looks to building traditions developed over centuries.[14] Heritage buildings provide answers for life within the tropics with minimal consumption of electricity. Natural ventilation, shade, and massing limit the need for air-conditioned spaces. Old streets and buildings can point the way, but not if they no longer exist.

Kerala is not a land of grand monuments, but it does have a fabric of human-scaled buildings interspersed with abundant nature. The need is for conservation programs that look at the vernacular buildings and landscape with a goal of honouring their ingenuity. The elite standards of UNESCO, the Venice Charter, and the ASI seldom accommodate vernacular environments, especially those of the working poor. It is time to create new models for conservation. The following ideas are particularly relevant for Kerala:

1) *A city architect appointed for communities with large historic enclaves.* This professional could advise and assist property owners. He/she should not police the community but act as an advocate and liaison for community members who wish to maintain their religious structures, homes, and shops. Citizens could be educated on building integrity and connected with local craftsmen/artisans.

2) *Designation of areas.* If only the monument (in some cases the central mosque or temple) is protected, there is danger that the community that supports that monument would be forced to relocate and the context of the area vanish, therefore weakening the relevance of the mosque itself. The designation must allow for growth, but it should be directed so that the heart of the city can continue to serve as market and home with new buildings inserted into the ensemble in such a way that all benefit. An agreed upon plan for conservation ensures that growth occurs in harmony with existing resources. This plan must reflect awareness of the delicate balance between the need for regulations and the need to adjust for the living conditions of existing communities.

3) *Grants to assist conservation.* The state government spends a substantial amount to encourage tourism, but allocates no funds for rehabilitation in old neighbourhoods. No incentives (such as tax credits) have been established for conservation of vernacular buildings. Parks, hotels, and waterfront improvements serve tourists, provide some local jobs, but do little to improve services for the local population. Tourists actually appreciate

a "living" city, full of residents at work and play. There are numerous ways to rehabilitate an urban neighbourhood without displacing the residents. In Europe, public housing money has been employed for renovation projects as well as for new construction. While maintenance of buildings is slow and unglamorous, it is also crucial. Building maintenance techniques should be demonstrated and encouraged – the known process of repairing roof tiles, whitewashing, or re-painting a building after the monsoon can have a lasting impact on the life of the building.

4) *An institute for traditional building arts training.* Kerala needs a centre for information on traditional materials and for training young people interested in working with timber, stone, and plaster. Knowledge of traditional building skills is in danger of disappearing with the death of older artisans. The conservation of vernacular buildings would not only invigorate the old city but also create much-needed new jobs for the young and ensure the ongoing continuance of age-old crafts. Kerala has a long tradition of public education and universal literacy. A building trade institute would be a fine addition.

Kochi and Kozhikode have the potential to become world heritage sites, "related to the spice trade between Asia, Arabia and Europe, in which the Malabar Coast plays a key role" (UNESCO 2011). But to succeed in maintaining the communities, leadership must turn away from focusing only on "monuments". The unique fabric of trading ports should be emphasized, including the mosque compounds. These anchor the historic neighbourhoods and act as starting points for conservation.

Equally important is a discussion on tourism in the old city. A living heritage requires the continued presence of residents. A balance between residents and tourists and the services required for each population is paramount. The growth of tourism in Kochi has eaten away at the Fort Cochin community and is now spreading down Bazar Road. "Heritage" is sold to the tourists without being clearly defined or integrated into the local consciousness. Speaking of Kochi, the authors of a 2013 conservation report could be speaking of much of the old urban fabric of Kerala:

> While old city areas are being frozen in conservation efforts, making them completely irrelevant in the lives of larger sections of the citizens, and being either made consumables for the tourism industry, or as 'museum' areas, zones which cannot sustain such 'freeze and preserve' approach are seeing constant and rapid erosion of its heritage environment, a total degeneration of heritage zones, and a loss of an immense wealth of heritage urban environment, being rapidly replaced by an energy guzzling, decontextualized globalized city.
>
> (c-hed, 29)

Not only are the mosques disappearing but so are their surroundings: the palm-treed properties, the outbuildings, the neighbouring homes and shops.

As Kerala and India embrace the 21st century, there is an urgent need for conservation education and training around preservation of vernacular historic structures.

Notes

1 Pepper, *piper nigum*, also known as "black gold", is native to Kerala and was the most important commodity of ancient trade. Other major items of trade were timber and textiles.
2 "...before the arrival of the Portuguese in the Indian Ocean in 1498 there had been no organized attempt by any political power to control the sea-lanes and the long-distance trade of Asia" (Chaudhuri 1985, 14).
3 2011 Indian census. www.censusindia.gov.in
4 Tales abound that the apostle St Thomas arrived on the Malabar Coast in his lifetime. Many Catholics consider themselves Syrian Christians or St Thomas Christians because they pay homage to the Church of the East. The St Thomas and Syrian Christians have a long history in Kerala and they remain a strong force in the community.
5 Many authors have undertaken studies on the integration of Islam into the Indian world. One example is the work of Richard Eaton, who states: "From the seventh century on, Muslims everywhere had been engaged in projects of cultural accommodation, appropriation, and assimilation, which had the effect of transforming what had begun as an Arab cult into what we call a world religion. ..." (2003, 4).
6 Kochi is probably the Kerala leader for reports. The Centre for Heritage, Environment and Development (c-hed), a department of the local government, has a library full of analysis, surveys and convocation reports that date back at least thirty years.
7 For more information on the dig, see *Unearthing Pattanam*, by P.J. Cherian and Jaya Menon, the catalogue for the 2014 exhibit at the National Museum, New Delhi.
8 Manas Roshan "History lost ...twice?" *Hindustan Times*, 1-28-2017. Numerous other articles chronicle other controversies about methods and conclusions. The dig has brought forth a great difference of opinion among Indian archaeologists.
9 Ibn Battuta wrote of the wealth and prestige of the Muslim merchant community in the 14th century. Since then countless scholars have studied the role of trade in the Indian Ocean and the interactions between Islamic and Hindu culture (for more see K.N. Chaudhuri 1985; Dale 1980; Abu-Lughod 1989; Gupta 1967).
10 India Pepper and Spice Trade Association (IPSTA) lists itself as the oldest commodity exchange in India (www.ipstaindia.com).
11 Mappila Muslims are descendants of local converts to Islam who married foreigners, often Arabs from the Hadramawt in Yemen, many generations ago. See also S.F. Dale (1980) and A.A. Engineer (1995).
12 A classic on conservation of old cities is Donald Appleyard's *The Conservation of European Cities* (1979). Examples abound of older buildings rehabilitated into vibrant contemporary centers, conserving all of the medieval structures. Closer to Kerala in climate and materials, the work of the Penang Heritage Trust in updating the city center's shophouses can be found at http://www.pht.org. my. A heroic effort has been undertaken by local conservationists to maintain the existing community in their workshops and homes and not fall victim to the commodification of the city. See www.gtwhi.com.my for work in George Town, the UNESCO World Heritage site in Penang Malaysia.

13 See "Mishkal: A Mosque that withstood centuries". *Malayalam*, 6-19-2011 and Amitav Rajan "PSU bends rules to fund temple and mosque to keep PMO happy." *The Indian Express*, 9-9-2008.
14 "earth's fossil fuels are a finite resource...no one can any longer afford to demolish buildings that are basically sound" (Cantacuzino 1985, 21).

References

Abu-Lughod, Janet. 1989. *Before European Hegemony*. New York: Oxford University Press.

Appleyard, Donald, ed. 1979. *The Conservation of European Cities*. Cambridge: MIT Press.

Cantacuzino, Sherban, ed. 1985. *Architecture in Continuity, Building in the Islamic World Today*. New York: Aperture.

Centre for Heritage, Environment and Development (c-hed). 2013. *Project to Revamp the Historic Trading Area and Cultural Landscape around Broadway in Cochin*. Kochi: Centre for Environment and Development.

Chaudhuri, K.N. 1985. *Trade and Civilisation in the Indian Ocean*. Cambridge: Cambridge University Press.

Cherian, P.J. 2014. Unearthing Pattanam. Catalogue for the 2014 exhibition at the National Museum, New Delhi.

Correa, Charles. 2012. *A Place in the Shade*. Germany: Hatje Cantz Verlag.

Dale, Stephen Frederic. 1980. *Islamic Society on the South Asian Frontier. The Mappilas of Malabar*. Oxford: Clarendon Press.

Das Gupta, Ashin. 1967. *Malabar in Asian Trade 1740–1800*. Cambridge: Cambridge University Press.

Desai, Miki. 2003. "Wooden Architecture of Kerala: The Physicality and the Spirituality". *Traditional and Vernacular Architecture*. Chennai: Chennai Craft Foundation.

Eaton, Richard M., ed. 2003. *India's Islamic Traditions, 711-1750*. New Delhi: Oxford University Press.

Engineer, Asghar Ali, ed. 1995. *Kerala Muslims A Historical Perspective*. New Delhi: Ajanta Publications.

Fels, Patricia Tusa. 2011. *Mosques of Cochin*. Ahmedabad: Mapin Publishing.

Firoz, Ar. Mohd and Students. 2009. *Kuttichira Settlement Revitalization and Conservation*. Calicut: National Institute of Technology.

Kasthurba, A.K. 2012. *Kuttichira – A Medieval Muslim Settlement of Kerala*. Calicut: Malabar Institute for Research and Development.

Krishnakumar, Vaisali. 2014. *Conservation of Kuttichira*. Unpublished Thesis, School of Planning and Architecture, New Delhi.

Manthanah, Aneesh. 2015. *Adaptive Reuse of Koya House*. Unpublished Thesis, National Institute of Technology, Calicut.

Meribha, Adline. 2014. *Adaptive Reuse of Tharavads in Kuttichira*. Calicut: INTACH.

Metcalf, Barbara D. and Thomas R. Metcalf. 2002. *A Concise History of India*. Cambridge: Cambridge University Press.

Mirchandani, Maya. 2015. "In Search of Muziris: India's Biggest Archaeological Dig". *NDTV*, August 30.

202 *Patricia Tusa Fels*

Riello, Giorgio Riello and Tirthankar Roy, eds. 2009. *How India Clothed the World*. Leiden: Brill.

Shokoohy, Mehrdad. 2003. *Muslim Architecture of South India*. London: RoutledgeCurzon.

UNESCO New Delhi Office. 2009. *Conservation and Revitalization Plan for the conservation Zone of Fort Cochin and Mattancherry*. New Delhi: UNESCO.

UNESCO New Delhi Office. 2011. *Conservation and Development Plan Project Report Phase 2*. New Delhi: UNESCO.

13 Craft as intangible heritage

The Thatheras of Jandiala Guru, Punjab

Yaaminey Mubayi

Introduction

The craft of copper and brass utensil-making practised by the Thatheras of Jandiala Guru, Punjab was inscribed on UNESCO's Intangible Cultural Heritage (henceforth ICH) list in 2014. This is the first time that a traditional craft form from India, metalwork, has been placed on the UNESCO list. It is anticipated to open the doors to ward recognizing the cultural value of craft and its practitioners and highlighting their significance as the "living heritage" of India.

The significance of the Jandiala Thatheras lies not in their uniqueness: there are Thathera settlements all over the country and across Afghanistan, Iran, Turkey, Egypt, and Morocco. Their uniqueness is in the manufacturing process, drawing from a complex range of traditional knowledge systems including medicinal, ritual, and technological, combined with the intense involvement of the craftsman with every piece emerging from his hands, embodied in the finished product. This process is handed down through the generations by a hereditary patriarchal transmission of knowledge, predominantly from father to son. Women play a marginal role in the manufacturing process, as they are considered to not have the physical strength to wield the heavy hammer, the primary tool of the Thatheras. The multi-dimensionality of the technique makes it extremely difficult to replicate by mechanical manufacturing processes. The inscription by UNESCO celebrates this skill as the human context of the product, encouraging a deeper understanding of what makes a craft and its products living history. The products created by the Thatheras are manufactured using traditional metalworking skills that comprise a knowledge base developed over generations. These skills are recognized and celebrated in the UNESCO inscription, and are what make the craft and craftsmen living history.

The inscription of the Thatheras on the UNESCO list is a significant move, one that addresses several aspects of craft conservation in India. At the core of the inscription is a recognition of a livelihood, complete with its cultural as well as commercial dimensions, as a form of India's intangible

Figure 13.1 A traditional Thathera craftsman at work. Source: Yaaminey Mubayi

heritage. This marks a paradigm shift from the more general perception of visual and performing arts, ritual and symbolic traditions as constituting a society's intangible cultural heritage. The inscription will have a lasting impact on future policy and attitudes towards the cultural value of handicraft in India, its markets and sustainability.

The first section of this chapter examines the evolution of ICH as a normative category within UNESCO's expanding mandate for the Culture Sector[1] over the past few decades. The second part examines the nomination process itself as a dialectical interaction with the practitioners, delineating the traditional practice within the transforming context of their historic past and the changing present, and explores possibilities for a sustainable future. The final section details the pedagogical implications of craft documentation for young professionals working in the field of heritage conservation (Figure 13.1).

The ICH Convention of 2003 – valuing intangible heritage[2]

The conventions relating to culture adopted by UNESCO over the past six decades reflect the need of member states to have international norms and best practices that augment and shape national development policies. To that end, UNESCO's normative instruments in the field of cultural heritage have evolved with changing perceptions of the role of culture in the development of societies. In the post-WWII decades, UNESCO's early policies and strategies in the field of culture were aimed at understanding emerging national and ethnic identities in the context of decolonization. The domains of music, literature, museums, and languages were at the forefront of these initiatives that led to the production of *Unity and Diversity of Cultures* in 1953, a publication series based on extensive surveys of societies and communities around the world (UNESCO, 1953).

In 1966, the Declaration on the Principles of International Cultural Cooperation was adopted by UNESCO, recognizing that all cultures have dignity and value and communities have the right to practice their cultural traditions. Moreover, the varied cultures of diverse communities collectively form the common heritage of mankind. Initiatives such as the Nubia Campaign wherein the Abu Simbel temple in Egypt was saved through international cooperation, led the way in formulating the convention concerning the Protection of the World Cultural and Natural Heritage (The World Heritage Convention) in 1972. Although this convention focused on built heritage, it strengthened the idea of a common heritage of humanity, and the onus of member states to protect it. It also introduced a programmatic approach based on a listing system and the use of operational guidelines for implementation.

Following the success of the World Heritage Convention, and in conjunction with the growing movement towards the recognition and protection of intellectual property rights, experts advocated that the scope of culture be extended to include values, worldviews, and beliefs of communities. The World Conference on Cultural Policies held in Mexico in 1982 to review knowledge and experience gained on cultural policies and practices advocated that attention be given to the preservation of "Intangible Heritage". It also recognized "creative expressions" as an intrinsic aspect of Cultural Heritage. In 1989, in Hammamet, Tunisia, a meeting of experts was held to formulate a ten-year plan for safeguarding handicrafts. Over the following decade, UNESCO carried out extensive inventories of craft, folklore, and other forms of intangible heritage. In 1992, the agency published *Our Creative Diversity*, highlighting the value of heritage transmitted through the generations, and recognizing that physical objects like crafts as powerful representations of the intangible heritage of communities. The report also stressed important issues like authenticity and fear of commodification, both applicable to the craft sector (UNESCO 1995).

In 2001, the Proclamation for the Preservation of Masterpieces of Oral and Intangible Heritage was promulgated for safeguarding intangible heritage. Finally, in 2003, the Convention for the Safeguarding of the Intangible Cultural Heritage was adopted. Its structure is based on the programmatic approach followed by the World Heritage Convention of 1972, but unlike the former, the 2003 convention does not operate through the notion of "outstanding universal value", but regards all forms of ICH as equally significant. Its main aim is to ensure visibility of cultural expressions and also their transmission through the generations as a form of preservation. The 2003 Convention is complemented by the Convention on the Protection and Promotion of the Diversity of Cultural Expression in 2005, which deals with the production of cultural expressions, and their circulation through activities, goods, and services. The process for nomination of diverse cultural expressions is premised on the principles advocated by both conventions.

The nomination process: "discovering" the Thatheras' history, practices, and issues

The proposal for nomination of the Thatheras of Jandiala Guru onto the UNESCO ICH list emerged from a wider historical documentation exercise that was carried out in Punjab in 2008–09. I had observed the potential of the Thatheras as contenders for the UNESCO ICH list, given their well-established position within the craft networks of Punjab, and their historic role in the exchange of skilled communities during India's Partition in 1947. When I suggested their nomination to the Department of Tourism and Culture, Government of Punjab, in 2009, I was asked to prepare the documents which were then submitted to the nodal agency of the Ministry of Culture, Government of India, and were then forwarded to UNESCO as one of India's entries for that year. Following several iterations, the craft of the Thatheras was finally inscribed on the ICH in 2014.

The inscription provides visibility to the craft, and advocates guidelines for maintaining the authenticity of the manufacturing process and the product. Unlike the case of built heritage, intangible heritage remains susceptible to loss of knowledge systems with the passing of successive generations. It is extremely vulnerable to changes in the manufacturing ecosystem, i.e. dwindling of raw material supply, changes in the taxation structure, even evolving consumer tastes and preferences. Hence the ICH listing process strongly stresses the documentation of traditional skills and knowledge systems and their historic and ethnographic significance, so that technological changes may be accommodated within an overarching authenticity so the "spirit" of the craft is sustained.

Thatheras: background and history

It is important to step back and observe the geo-political significance of the Thathera settlement of Jandiala Guru as it lies within the historic commercial hinterland of Amritsar. The latter has traditionally been known as the "Mecca of the Sikhs", the overarching presence of the Golden Temple being a dominant factor in its evolution through history. However, the city has another identity, that of a significant trading zone and commercial centre, that has had a major role in its development. It is not a coincidence that Amritsar is situated on the major overland trade route of northern India, the Grand Trunk road, that interweaves with a network of routes connecting with the Silk Route at Kabul. Its geo-political significance was recognized by Sri Guru Ramdas, who had excavated the sacred *sarovar* (tank) in 1573. He also simultaneously invited members of fifty-two crafts and trading castes and established a working community of skilled artisans and craftsmen near Sri Harimandir Sahib. This first settlement of thirty-two shops called Battis Hatta, is still in existence. Thus, a link between faith and livelihoods was forever established that ensured the survival of the city despite the turbulent history of raids and attacks on Sri Harimandir (Grewal 2004).

The reign of Maharaja Ranjit Singh (1780–1839), founder of the Sikh kingdom, may be termed a "renaissance" of art and craft forms in Punjab, when patronage was extended to all types of hand-crafted products. In the 1830s, Amritsar and its environs received a large influx of craftsmen from Kashmir and Rajasthan, and trade and manufacturing of textiles, embroidery, jewellery, carpet weaving, metalwork, woodcraft, etc. flourished as markets for the products boomed. During Maharaja Ranjit Singh's reign, the city became vitally linked with the monarch's political capital at Lahore, which continued to provide a market for Amritsar's trade and manufacturing activities through the colonial period until 1947 (Gauba 1988).

Jandiala Guru is a small town on the Grand Trunk Road, about ten kilometres from Amritsar city in Punjab. It was founded in the early 19th century by Jats (a caste of landowning agriculturists) and named after Jand, the son of the founder (*Gazetteer of Amritsar District 1883–84* 2000). The original town was fortified by a mud wall with seven gates. Soon after, it was settled by a community of metal craftsmen, likely from Kashmir, who specialized in manufacturing copper, brass, and *kansa* (an alloy of copper, tin and zinc) utensils (Amritsar District Gazetteer 1883–84). During the Partition of India in 1947, the Kashmiri Muslim craftsmen migrated to Pakistan, while their Hindu counterparts from across the border in Kujranwala, moved to Jandiala. (The present settlement consists of about 400 households of Thatheras, migrants from Kujranwala.) However, memories of the original craftsmen live on in the settlement's name, "Bazaar Thatherian, Gali Kashmirian". Even before Partition, there was contact and exchange between Kujranwala and Jandiala, as two *mandis* (market centres) of metal utensils. Despite the change in population, Jandiala Guru remains a crafts settlement, with its traditional craft intact.

The Thatheras belong to the Khatri caste, and consider themselves distinct from the Jats who settled Jandiala in the 19th century. The Thatheras proudly declare that they do not wield the plough (*Hal*) but the hammer (*Hathoda*), thereby embracing their identity as craftsmen in an urban setting. In modern administrative terms, Jandiala is a small commercial town (*qasba*) with a municipality (not a *Gram Panchayat*).

Figure 13.2 View of a *bhatti* – traditional oven. Source: Yaaminey Mubayi

In terms of the built environment, the manufacturing settlement is laid out along a series of narrow lanes, lined on both sides by small workshops and sheds, each having an adjacent residential space for the families of the craftsmen. In fact, the intermingling of work and family life is a common practice among traditional craftsmen in the subcontinent, and is exemplified by the clay oven (*bhatti*; Figure 13.2), used for both heating the metal and cooking lentils (*dal*). The sound of heavy wooden hammers striking metal is a constant refrain as one walks through the streets. The neighbourhood's built fabric clearly reflects its historic origins through architectural elements like carved wooden balconies, door and window frames and the occasional extant wall constructed with narrow *nanakshahi* bricks set in mud plaster.

Process and products

Utensils made of brass, copper, *kansa* (copper, zinc, and tin alloy) are the traditional products of Jandiala Guru. These range from small bowls (*katora*), rimmed plates (*thali*) to larger pots for water and milk (*gagar*) and huge cooking vessels (*degh, pateela, karahi*). While the smaller vessels are used in individual households, the larger ones are used in big community kitchens at temples, gurudwaras, and wedding banquets.

A rolling mill is situated at one end of the settlement. Metal scrap is melted here in big underground furnaces, poured into iron moulds and

Figure 13.3 View of the urban form of Jandiala Guru. Source: Yaaminey Mubayi

cooled. The cooled "cakes" of metal are then passed through mechanized rollers and flattened into thin plates (Figure 13.3). These plates are then bought by the Thatheras, who hammer them into curved shapes, welding them together to produce the required pots, urns, plates, and bowls. The individual workshops of the Thatheras also have large mud-brick kilns and can occasionally melt the scrap themselves.

Heating the plates while hammering and curving into different shapes requires careful temperature control, and this is done on tiny stoves buried in the earth and fired by wood chips. Handheld bellows help to increase or reduce the intensity of the flame and the joints and rivets are strengthened by alternatively heating and cooling (Figure 13.4). All edges are hammered to smoothness while the metal is still hot, and then cooled with water to seal them. The utensils are "finished" by polishing them with dilute hydrochloric acid, sand, and tamarind juice, all done by hand. Designs are made on the vessels by skilfully hammering a series of tiny dents on the heated metal. The skill and detail of the design varies from region to region – the jugs and samovars of Kashmir are beautifully carved with intricate paisley patterns not unlike the paisley (*jamawar*) shawls or the copper pots of Almora featuring chains of floral designs. The vessels in Jandiala have a more basic utilitarian appearance, embellished with regular series of dents and dimples.

Figure 13.4 Rolling mill manufacturing metal plates. Source: Yaaminey Mubayi

Ethnographic significance of the craft and its products[3]

Thatheras generally live in patriarchal family units, and the male members work the family trade. Occasionally, help from neighbouring villages is employed, when there is seasonal pressure of work, for instance, during the wedding season in February–March, or during the annual fairs in October. As *Khatris*, a non-agricultural caste, Thatheras are considered higher on the caste hierarchy in North Indian society. Their products play an important role in social functioning and demarcating rites of passage and other domestic milestones. This throws light on the fact that traditional crafts of India are not simply random products to be used and discarded. Each item carries a unique symbolism, associations with rituals, family, and rites of passage.

The Thatheras' vessels were a part of intimate family rituals and customs in earlier times, especially those pertaining to birth, marriage, and death. Should a baby's upper teeth emerge before the lower ones, an event considered unlucky, the teeth are tapped with a *Jagannathi*, a *kansa* (bronze) bowl, by the child's *Nani*, the maternal grandmother, who also gives gifts to the child. When a son is born after three daughters, it is considered unlucky, and the newborn is passed through a hole specially made in the roof of the house. Alternatively, the rim of a *kansa thali*, with the base removed, called *Trikhal*, can be used to pass the baby through. This signifies a new birth for the child and the father can only view his son after this ceremony. At the time of a marriage, the bridegroom is fed curd (yogurt) and honey in a bowl called *Madhupurkha* following the nuptials. A bride and bridegroom viewing one another for the first time after marriage is reflected in a *Chhayapatra*, a flat dish in which some oil is poured. A brass *paraat* (flat vessel with high sides) is used for gathering and washing the bones of the dead after cremation, after which it is donated along with the *lota* (water pitcher). Temple *arati* (congregational singing) is accompanied by a *vijay kanth*, a brass gong.

Thus, the Thatheras, via their specialized vessels, enable the interweaving of a large network of social relations, a factor that contributes to their integration with the fabric of society. This intertwining of social- and material-, or livelihood-related aspects is characteristic of traditional economies in India.

Marketing and legal regime[4]

Jandiala Guru has been a well-known historic manufacturing centre and a wholesale market for brass and copper utensils (*mandi*) for almost two hundred years. Until a few decades ago, there were several such *mandis* that had flourished in north India. In Punjab, there were many other such centres including Batala, Hoshiarpur, Phagwara, and Jagadhri. These *mandis* survived on the basis of fixed areas of demand that assured their sustainability through dependable markets. The traditional market economy for brass and copper utensils operated through the medium of buyers from the market centres placing orders and purchasing in bulk, thereby assuring adequate credit for raw material and capital to roll over for production and profit. The network of traditional markets was a far-flung one, crossing regional boundaries and linking diverse ethnic and linguistic communities through the medium of trans-regional trade.

The utensils produced in Jandiala Guru have historically found markets in various parts of north India. Brass vessels with high sides and tight conical lids, called *Rangjhan*, are used in the monasteries of Ladakh for cooking rice. Large *deghs* (deep open-mouthed pots) are used for cooking on a large scale at fairs and festivals in Himachal Pradesh, for instance, at the traditional fair in Rampur-Bushehr and the Dussehra fair in Kullu. *Deghs* are also in demand as dowry items among Muslim families in Najibabad, Uttar Pradesh. Large brass *karahis* (round-based frying vessels) are a favoured item among the Rah Sikhs of Ferozepur and Jalalabad (Fazilka area). In general, big cooking vessels find a ready market in Gurudwaras that cater to large-scale cooking for a *langar* (community kitchen).

Issues

Despite the ubiquity of Thathera settlements across northern India, the community along with its traditional skills is an endangered one. The problems faced by the Thatheras of Jandiala Guru must be seen in light of larger political and social issues besetting Punjab as a state and the country's small industrial sector.

Despite being considered a developed state with a high per capita income, the problems faced by Punjab are many and include over-dependence on agriculture and lack of industrial development, unemployment and underemployment, and demoralization of the youth, among others. These have a socio-cultural dimension, apart from an economic one. Following the Partition of India in 1947, the Amritsar region, which was dependent on Lahore as a market for its products, found itself cut off from it and hence went into a commercial and industrial decline. The entire pattern of industry changed in the country, with industrial centres developing in Maharashtra, Gujarat, and southern India. Owing to its proximity to the national border, the Amritsar region was neglected in terms of both industrial investment as well as civil society initiatives (Government of India 2012). There are relatively few civil society organisations in terms of self-help groups and local cooperatives in Punjab, contributing to the vulnerability of craftsmen to market forces.

The main problems faced by the Thatheras of Jandiala Guru include lack of essential services like power supply, poor working conditions, an unfavourable taxation policy environment[5], dwindling public and private support in terms of raw materials, and a general disempowerment of the craftsmen (Ashok Mehrotra, Secretary, Thathera Handmade Utensil Association, in discussion with the author). For example, the introduction of the Value Added Tax (VAT) in 2005 and the Goods and Services Tax (GST) in 2017 has had a disastrous impact on the Thatheras' retail, as they simply do not have the revenue levels or purchasing power to pay the extra taxes. This is leading to a diminution of the craft with the younger generation seeing no future in learning and following the craft of their forefathers. The dwindling market demand, too, indicates that the traditional products are being devalued in the open market where their value as a community's heritage is not considered, and where, in terms of economies of scale, they cannot hope to compete with the mass-produced objects manufactured in large factories.

The problems specifically faced by the Thatheras of Jandiala Guru may be categorized into two broad areas. In the first instance, the community is challenged by an unfriendly and unfamiliar legal and taxation structure, that does not enable their craft but marginalizes them from the mainstream narratives of national development. This includes the fact that sales tax and VAT are applicable to their products, but owing to their small scale, it makes their products expensive and unviable in the traditional markets. In the 1950s and 60s, they received some subsidies on coal and scrap metal from the government, especially when the Indian Army gave them their shell casings. However, these subsidies have since been discontinued, and the Thatheras have to buy scrap metal in the open market, competing with mainstream industries in the sector.

Owing to the fact that they are not organized into either cooperatives or private limited enterprises, but function largely as individual entrepreneurs, they have not benefited from the various government schemes, such as the Handlooms Act of 1994 or the PMEGP 2008 (Prime Minister's Employment Generation Programme). They lack both organization and capacity to apply for information about the schemes. Their lack of organization also prevents them from lobbying for better working facilities and better power supply with the state and local governments. The result of their disempowerment is that they are increasingly abandoning their traditional products and processes and opting to produce cheaper and mass-produced merchandise like aluminium vessels, for which there is a larger market. More importantly, their children are abandoning the family skills and professions, opting for easier and more marketable vocations like business, commerce and information technology. The skills of the Thatheras are truly on the decline.

The Thatheras of Jandiala Guru are the last of the living metal workers' *mandis* in Punjab. As one Thathera wistfully recalled: "when we were children, we used to take pride in learning the craft from our fathers and uncles. Now our children do not want to wield the hammer, they want computer jobs. No one wants to work with their hands anymore. Our skill will die with us" (Ashok Mehrotra, Secretary, Thathera Handmade Utensil Association, in discussion with the author). And with the craft, its knowledge systems, values, and work ethic will also pass into oblivion.

Reversing the trend – possible solutions

The significance of the Thatheras lies not in their uniqueness, but their ubiquity. Historically, they have filled a need within the society and economy of communities and are networked into the cultural fabric of the region. The UNESCO inscription acknowledges the link between the process of manufacture of the products and its social and cultural context. The inscription recognizes the historic identity of the Thatheras in conjunction with their way of life, rhythms of work, and focuses their identity on their craft, their greatest asset.

At the community level, the next step would be the organization of the craftsmen into a working collective. At the moment, the Thatheras' Association is virtually defunct, and each craftsman is his neighbour's competitor, attempting to promote his business by undercutting the others. The collective identity needs to be revitalized so that entire community can find a common voice, and problems and benefits can be shared.

A collective organization can be incubated and supported by the district administration through capacity-building agencies like the District Industries Centre (DIC), which can link them into a network of public schemes involving design development, training, and marketing. Their products have already been included by DIC on the list of potential merchandise that can be taken up for capacity-building. However, the inscription by UNESCO would give the process much needed momentum and a greater priority. For example, currently, the brass utensils of Jandiala Guru are being considered as a product to be showcased in various heritage tourism initiatives of the Department of Tourism, Government of Punjab.

At a state and national level, the inscription could lead to an acknowledgement of the craft as a skill practiced by master craftsmen. It would enable the identification of the product and the manufacturing process along with its community and the region, leading to the creation of a brand: the brass and copper utensils of Jandiala Guru in Punjab. This could then be marketed through the electronic media, highlighting the product, practitioners, and the process in a holistic manner.

The recognition of the craft, its practitioners, their working and living spaces, and the process of manufacture as a consolidated product can also have positive implications for the development of tourism. Its inscription by UNESCO placed it onto an international tourism map, bringing the entire complex into the limelight as a tourism product. It could be included as a destination for various tourism initiatives being considered in the region by the state and national governments. This would open the door to upgradation of municipal services, roads and connectivity, power supply and enable innovations in upgrading working conditions, training and capacity-building of the practitioners, particularly the younger generation.

The implementation of the UNESCO ICH convention in India holds tremendous potential for the survival and sustainability of the country's traditional knowledge systems. India has ratified both the Conventions of 2003 and 2005, and has shown enthusiastic commitment to identifying and preserving her intangible heritage. However, its implementation within the country has not been without challenges. Unlike the case of the World Heritage Convention, wherein the mandate for implementation of the international norms was smoothly adopted at the national level by the Archaeological Survey of India, there was no "ready-made" agency with the mandate and capacities for implementing the ICH Conventions. Also, unlike the field of built heritage which is located within a single ministry, the Ministry of Culture, ICH consists of multiple disciplines that are scattered across diverse departments and ministries of the Government of India. For example, handicrafts are part of the Ministry of Textiles, while the craft of the Thatheras is located within the Department of Industries of the state government. A satisfactory operational structure to override the various bureaucratic boundaries and allow the suitable implementation of the internationally accepted norms, is yet to evolve.

The inscription by UNESCO would enable the establishment of the craftsmen's collective organization, a crucial element to protect the traditional craft from social and cultural erosion by misguided commercial development. It would enable a constructive dialogue to be generated between the community and the state. By bringing the product and its related issues into the limelight, it would enable access to international and national expertise and best practices. The recognition of the craft and its context as the community's collective heritage would contribute tremendously to ensuring its continuity and development in a sustainable manner.

The establishment of local-global linkages is vital for the craftsmen to find voice and empowerment. The free flow of information, whether through the internet or enabling laws like the Right to Information Act, 2005 would ensure at one level access to market information, development schemes like the Seal of Excellence for master craftsmen,[6] and training and educational programmes for the youth. At another level, it would enable transparency in implementation, doing away with disempowerment through lack of knowledge of rights, entitlements, and procedures. It would remove the monopolistic exploitation of individual craftsmen by middlemen and vested interests. Finally, it would enable the national and international markets, development professionals and institutions to gain access to the craft of Jandiala Guru, which would no longer be an unknown pocket of rural Punjab.

UNESCO's ICH nomination process as a pedagogical tool – lessons from the field

It is an oft-repeated notion that cultural heritage is an under-theorized field, lacking conceptual frameworks that take it beyond empirical learning into a more abstract disciplinary understanding of the sector. Heritage is also a fractured intellectual domain in India, with major gaps between the architecture-based realm of "built heritage" and the multifaceted domain of intangible heritage with its sociological and anthropological practices. The latter is even more vulnerable from a disciplinary perspective than the former, requiring a more rigorous formulation and application of methodologies to achieve greater professional credibility. The vulnerability factor is enhanced by the reality that several aspects of intangible heritage, particularly craft, are directly related to human livelihoods, and the close alignment of theory with practice is essential as it powerfully impacts people's lives and well-being. At the moment, the policy space is not adequately informed either by conceptual understanding or experiential knowledge of the domain, leading to unsatisfactory policies that do not address the complex dimensions of the field. It is necessary, therefore, to create a cadre of heritage professionals possessing theoretical expertise coupled with empirical insights who can generate the required interface.

To bridge the gap between theory and practice, the ICH listing process offers a suitable point of entry for heritage professionals. The interdisciplinary nature of the nomination process, encompassing historical, ethnographic, technical and developmental research, generates an appropriate platform for addressing the multi-dimensional issues at stake. The preparation of an action plan and roadmap is premised upon a deep understanding of the domain and the community. It is a rigorous and action-oriented methodology that goes into generating an implementable and beneficial policy (Foster 2011).[7]

The representation of the community and its context, historical and environmental, the production process, and the ritual and ethnographic significance of the products, in a short video format[8] is a powerful resource for the document. Preparing the film requires a fair amount of skill, rigour, and clarity in summarizing the subject and highlighting significant issues.

In conclusion, the inscription of the Thatheras of Jandiala Guru on the UNESCO ICH list has opened a window to the recognition of craft as cultural heritage in India. Its further advancement by evolving sustainable livelihoods for practitioners through design development and training, marketing, and tourism is yet to attain discernible targets. However, the first step, that of identifying the heritage value of craft, has been successful.

Acknowledgements

I wish to acknowledge the UNESCO ICH list, under which much of the information in this article is secured. The purpose of this paper is to record my experience in "discovering" and nominating this craft form for inclusion within the UNESCO list. I am immensely grateful to the State Government of Punjab, particularly the Punjab Heritage Tourism and Promotion Board, for their encouragement in enabling the nomination. I would like to thank Mr. D.P. Bhagat, General Manager, District Industries Centre, Amritsar for taking a personal interest in the project. Professor Molly Kaushal and her team at IGNCA, as well as Ms. Ritu Sethi of Crafts Revival Trust, helped move the nomination through the various stages towards completion. Above all, my deepest thanks are to the Thathera community of Jandiala Guru, for their warmth, openness, and generosity in sharing their skill and experiences with me.

Notes

1 Culture is one of the four operational areas of UNESCO, the other three being Education, Science and Communication. UNESCO is the only multi-lateral agency with a specific mandate to address the sector of Culture in its member states.
2 Barbara Kirschenblatt-Gimblett, "Intangible Heritage as Metacultural Production" http://www.nyu.edu/classes/bkg/web/heritage_MI.pdf
3 Information for this section was gathered through an interview with Shri Bachan Lal Thathera conducted by the author.
4 Information for this section was collected through an interview with Shri Ashok Mehrotra, member of the Thathera Handmade Utensil Association, Jandiala Guru. Interview conducted by the author.
5 The Value Added Tax (VAT) introduced in India in 2005, placed an additional burden on the unorganised sector, to which craftsmen like the Thatheras belong. As their supply chains were not formally and clearly defined, there was a great scope for misuse of authority by local police and tax agencies.
6 The UNESCO-AHPADA Seal of Excellence for Handicraft Products in Southeast Asia, instituted in 2001, is awarded jointly by the United Nations Educational Scientific and Cultural Organization (UNESCO) and the ASEAN Handicraft Promotion and Development Association (AHPADA)

7 The listing process enables the emergence of a new discourse premised on the critical evaluation of their own practice by the communities involved. This helps practitioners to chart the future of their craft while cognizant of its value as heritage. It also enables contemporary skills and marketing practices, like information technology, to come into play within the framework of traditional processes.
8 This is an essential part of the nomination dossier.

References

Foster, Michael Dylan. 2011. "The UNESCO Effect: Confidence, Defamiliarization, and a New Element in the Discourse on a Japanese Island". *Journal of Folklore Research*, 48(1): 63–107.

Gauba, Anand. 1988. *Amritsar: A Study in Urban History, 1840–1947*. ABS Publications, Amritsar.

Gazetteer of Amritsar District, 1883–84. 2000. Sang-e-Meel Publishers, Lahore.

Government of India. 2012. *Brief Industrial Profile of Amritsar District*. http://dcmsme.gov.in/dips/Amritsar.pdf

Grewal, J.S. 2004. *Social and Cultural History of the Punjab*. Manohar Publishers, New Delhi.

Kirshenblatt-Gimblett, Barbara. 2004. "Intangible Heritage as Metacultural Production". *Museum International*, 56(221): 52–65. doi:10.1111/j.1350-0775.2004.00458.x.

UNESCO. 1953. *Unity and Diversity of Cultures*. Imprint, Paris.

UNESCO. 1995. "Our Creative Diversity". *Report of the World Commission of Culture and Development*. Paris.

Part IV

Sustainable approaches to heritage conservation

14 Ghats on the Ganga in Varanasi
A sustainable approach to landscape conservation

Amita Sinha

Introduction

At Varanasi, where the Ganga reverses its flow, the 6.8 km riverfront is an iconic image of the city (Figure 14.1). Its built fabric consisting of *ghats* (steps and landings) descending from the historic palaces and temples to the river evolved over 800 years from self-organized systems of worship and pilgrimage. This cultural landscape is complex in its layering and detail and was resilient in its recovery from natural disasters as well as cultural upheavals. It is constituted by situated events, both natural – flooding, silting, and the changing flow of the Ganga – and cultural – ritual activities and performances in diurnal and seasonal rhythms tied to the river. The riverfront is a complex ecosystem in which spatial practices have evolved over centuries to respond to natural phenomena and its cycles and reflect Hindu beliefs about cosmic order. Currently the ecosystem is stressed by overuse and unprecedented levels of pollution in the Ganga, posing a challenge to the idea of the river as an archetypal symbol of purity and to the continuity of spatial practices. Patrick Geddes' town planning reports in early twentieth century India are a useful precedent in linking ecological planning with traditional cultural practices. In the systems approach, symbolic meanings of nature in traditional beliefs and practices are augmented with utilitarian functions in local energy cycles linking sun, flora, and fauna. Design strategies are proposed within this framework for sustainable conservation of the *ghats*, to preserve their cultural heritage and promote a healthy and resilient cultural landscape. These strategies, when implemented, will support cultural traditions, address pollution, and increase the capacity of the landscape to recover from flood events.

View of nature as heritage

The cultural landscape of the Ganga Riverfront reflects the ancient Hindu worldview in which the occurrence of natural phenomena is celebrated as manifestation of *rita* or cosmic order. *Rita*, derived from the Sanskrit root of the verb *ra*, meaning 'to go' or 'to move', is believed to underlie

Figure 14.1 Skyline of Varanasi *ghats.* Source: Author

the movement of celestial and terrestrial phenomena, of solstices and equinoxes, of flowing rivers and vegetal growth, and the cycle of seasons. The diurnal and seasonal rhythms of natural phenomena express this eternal order that governs human life as well. Participation in rhythms of nature brings harmony and happiness and reaffirms the universal order in the traditional belief. Time is conceived as cyclical as in the birth and death of all living entities that are part of nature. It is celebrated in festivals tied to seasons and harvests, and in life cycle rituals.[1] These cultural practices are spatial in that they produce the landscape through human interactions with the natural phenomena (Sinha 2015, 43).

On the banks of Ganga in Varanasi, the sun and the river are central to spatial practices and are worshipped as transcendent divine entities made immanent through their material form and physical presence. As symbols of natural archetypes of fire and water, they are sources of energy that create and sustain all life. Ganga, deriving from the Sanskrit verb *gam* meaning 'to go', is the prime symbol of purity in her capacity to cleanse and purify through her flow. She is liquid *shakti* (energy) as Shiva's consort and as Ganga Ma: life-sustaining like mother's milk (Eck 2015, 240). Her descent to earth nurtures millions living in the plains of Northern India. As a divine goddess and a flowing river that purifies all that it touches, the Ganga's spiritual and phenomenal forms are mutually constituted. The sun, worshipped as *aditya*, generates life as a source of light and heat. Its movements in the sky following the winter and summer solstices, known as *uttarayana* and *dakshinayana* respectively, are an expression of *rita*, bringing forth the seasons. The moon's waxing and waning, symbolic of death and renewal, is the basis of the lunar calendar in which the month is divided into auspicious (*shukla-paksh*) and inauspicious (*krishna-paksh*) halves.

At the *ghats*, time, space, and cultural practices come together in commemorating divine nature and engaging with its material forms. Human activities synchronize with the dynamic natural phenomena, both affirming the transcendent cosmic order. Ganga liberates one from the cycle of death and rebirth upon cremation on her banks. Circadian rhythms are affirmed in daily worship of the sun at dawn in the ritual *namskar* (folding of hands and chanting sacred verses) and in the evening *aarti* (waving of lamps) to Ganga. Shrines to *Aditya*, the sun god on the western bank were located to align with the sun's position in the sky, marking *yantras* (triangles formed of visual axes), thereby inscribing a certain order in the landscape (Singh 1994). Most shrines are now lost, but a few remain, and their tanks are used for bathing on the auspicious occasion of *Makar Sankranti* (January 14) when the sun begins its northward journey. Festivals such as *Ganga Dusshera*, *Mahashivaratri*, *Kartik Purnima*, and *Deva Deepavali* are celebrated in accordance with the lunar calendar, when devotees take a dip in the holy waters and the *ghats* are lit with lamps (Eck 1982, 253). The festivals commemorate Shiva whose matted locks break the descent of Ganga's fall, the fortnight of the annual visit of ancestors to earth, and *lila* (deeds) of Vishnu *avatars* (incarnations), those of Ram and Narsingh.

Challenges at the Varanasi *ghats*

The built fabric of the *ghats* – its steps, landings, historic palaces, temples, and shrines – was shaped by cultural practices based upon the value assigned to being in harmony with natural rhythms and belief in purity of Ganga. The riverfront landscape evolved as a complex ecosystem in which culture and nature are in a symbiotic relationship (Ramakrishnan 2003). The ancient settlement was situated on the west bank on the Rajghat plateau at the confluence of Ganga and Varanasi. Behind the high ridge were lowland water bodies amidst forests that gradually disappeared as the city expanded southwards (Singh 1955, 15; Eck 1982, 48). Small settlements around inland lakes gave way to the sprawling city and the riverbank was transformed over time from a natural wooded area dotted with *ashrams* (hermitages) to the heavily built up urban edge today used for the cremation of 32,000 corpses every year and with 35 drains and sewers emptying untreated wastewater and sewage.[2] The *ghats* being the public commons of the city are intensively used for recreation and myriad other activities

– washing clothes and bathing buffaloes, open air defecation, transporting goods, and discharge of sullage.

The riverfront is severely stressed today with the core beliefs and practices beginning to lose their validity. As pollution mounts in the river, the Ganga's centrality as the ultimate symbol of purity is weakening in the Hindu imagination. Cultural practices valorizing Ganga's divinity are themselves generating waste that contributes to the local point source pollution (Conaway 2015).

There is a disconnect between ideology and reality, and a mismatch between the concepts of Ganga as pure (*shuddha*) and clean (*swatcha*). Priests at the *ghats* believe that Ganga will always be pure even though not clean, leading to lack of religious mobilization in state efforts to clean the Ganga (Alley 1998). Certain rituals such as *aarti* have become spectacles and are in danger of losing their meaning. In these ritual ceremonies attended by hundreds of devotees and tourists, songs paying homage to Ganga as a goddess who cleanses and purifies the material world feel hollow and inane in face of dirt on the *ghats* and the shoreline. The belief in Ganga's purity and regenerative powers are challenged by all too visible evidence of widespread environmental pollution. Diesel boats with low-grade engines and incomplete combustion of fuel plying on the river contribute to air pollution.[3] Many inland water bodies have been filled up and have become dumping grounds for waste leading to groundwater contamination. The Ganga in Varanasi, far from nourishing life, is a source of water-borne enteric disease for those who bathe in it and drink its waters.[4]

Efforts to control pollution in the Ganga have been unsuccessful in part because of the top-down approach in wastewater infrastructure projects constructed under the aegis of the Ganga Action Plan launched in 1986 by the Government of India.[5] Alley (2016) points out that sewage pumping stations and sewage treatment plants in Varanasi do not function to their full capacity because of intermittent electric supply and the state monopoly has led to corrupt practices that make the system even more inefficient. Ganga was declared as the 'National River' of India in 2009 when Central Government established the National River Ganga Basin Authority to tackle pollution at the regional level. The Namame Ganga Scheme announced in 2014 by the Government of India has recommended bioremediation and in situ treatment of wastewater flowing through the open drains in addition to constructing new sewage and effluent treatment plants. Bioremediation would require considerably less energy and has the potential to involve local communities. A combination of decentralized, bottom-up and top-down, and central approaches may have higher rate of success in tackling what is proving to be an intractable problem.

A sustainable approach to landscape conservation

Heritage conservation in South Asia has been largely a state enterprise and has continued the colonial legacy in preserving monumental buildings (Sinha

2017, 2). The idea of landscape as embodiment of heritage has yet to find wide acceptance in mainstream discourse or the field of conservation practice. New government initiatives such as HRIDAY and PRASAD for sacred and historic cities focus on developing their infrastructure.[6] HRIDAY (National Heritage City Development and Augmentation Yojana) scheme is for heritage cities while PRASAD (National Mission on Pilgrimage Rejuvenation and Spiritual Augmentation Drive) covers pilgrim cities. Varanasi comes under both schemes. But so far, only infrastructure improvements such as street widening and building of sewage treatment plants have been undertaken. No comprehensive plan for *ghat* management has been drawn up although Indian National Trust for Art and Cultural Heritage (INTACH) has worked on the restoration of historic palaces on two *ghats*. The government schemes do not focus on the landscape, the product of a culture's relationship with nature that sustains tangible and intangible forms of heritage. The Ganga riverfront in Varanasi remains neglected as the setting and subject of immensely significant spatial practices tied to sacred natural rhythms.

Patrick Geddes' bio-centric approach as demonstrated in his planning reports for cities in India in the early 20th century is a useful precedent to consider in developing an alternative and complementary framework for landscape conservation. While Geddes' report on Varanasi can no longer be traced, his work in other Indian cities exemplifies the importance of decentralization, civic responsibility, and collaborative planning (Tyrwhitt 1947). As his letters to friends show, Geddes understood that nature is sacred in Varanasi and new life emerges from destruction (Lannoy 1999, 600). In his diagnostic survey of historic cities, proposals for conservative surgery of decaying areas, and improving sanitation through sewage gardening, Geddes emphasized renewal and placed human needs above all (Khan 2011). He believed that planning for individual and societal development should occur by linking cultural insights and ecological niches.

A century later, the Geddesian approach is valid for planning the Ganga riverfront in Varanasi. The cultural heritage embodied by the *ghats* is at grave risk when the core belief in Ganga's purity cannot be sustained and traditional practices lose their meaning. The complex layering of historic and vernacular architecture in the *ghats* has resulted in a mosaic of built forms within which old monuments are interspersed with ordinary buildings constructed in the last few decades. Many historic palaces are in the process of being converted into luxury hotels catering to high-end tourism and are not the focus of state conservation efforts that will ensure public use. Vernacular style buildings built haphazardly in the last few decades encroach into the public space of the *ghats*. The *ghats* themselves are not perceived as worthy of preservation and their design grammar is not well understood. The octagonal and square platforms that enclose the steps, aediculae and niches on the landings, and semi-fixed structures such as wooden platforms and jute umbrellas, create a syncopated rhythm that lends an aesthetic character to the *ghats*. They are settings of spatial practices tied to seasonal and circadian rhythms structuring a dynamic perceptual landscape.

Stresses caused by pollution resulting from mechanized traffic and activities of millions of people makes it imperative that conservation go beyond historic structures and civic infrastructure projects, and address issues holistically in a system-based approach. In systems thinking landscapes are interconnected webs of spaces, events, and flows of energy, not a collection of discrete inert structures devoid of use. Sustainability is 'the quality of not being harmful to the environment or depleting natural resource, and thereby supporting long-term ecological balance'.[7] For a sustainable approach towards cleaning the Ganga and ensuring a healthy and resilient riverfront, the conservation focus should expand beyond the monument to the cultural landscape. It is imperative that the spiritual and phenomenal aspects of nature be coalesced in projective planning. The chasm between Ganga's divinity and her material properties can be bridged such that her transcendent powers continue to be immanent in her earthly flow. Sacred symbolism can be integrated with the utilitarian value of sun, flora, and fauna in developing strategies for managing the complex eco-cultural system of the *ghats*.

Flows of animals and people, ritual ceremonies, waste, flooding, and sun and shade on the *ghats* were mapped in site workshops in Varanasi (January 2014 and 2016) conducted by the author with undergraduate and graduate students in Landscape Architecture and Architecture at the University of Illinois at Urbana Champaign, USA. The mappings revealed the relationships between riverine ecology, spatial practices, cultural beliefs, and built forms thus creating the frame for systems thinking essential for understanding the cultural landscape of the *ghats* as a situated event. The following set of strategies developed within this framework could lead to sustainable conservation of tangible and intangible forms of heritage embodied by the *ghats*.[8]

Sun

The sun is worshipped daily and during festivals – especially during the festival *Lorarka Chauth*. *Lorark*, the 'trembling sun' is venerated at the shrine and *kund* (tank) close to Assi Ghat (Singh and Rana 2006, 109). This *ghat*, at the confluence of Ganga and Assi Nala is an auspicious site for bathing, especially on *Makar Sankranti* when the sun sets on its northward journey known as *uttarayan*. However, its energy remains untapped. In January and June, the coolest and warmest months respectively, between 85% to 92% of the *ghats* receive sunshine from 9 am to 2 pm. Solar energy can be harvested from this abundant sunshine in solar panels mounted on the vacant site south of Assi Ghat where the riverfront is not actively used. Given the erratic power supply from the grid in Varanasi, this renewable energy can be an alternative source of electricity to be used in powering boat engines, in charging stations for mobile devices, and lighting the *ghats* (Figure 14.2).

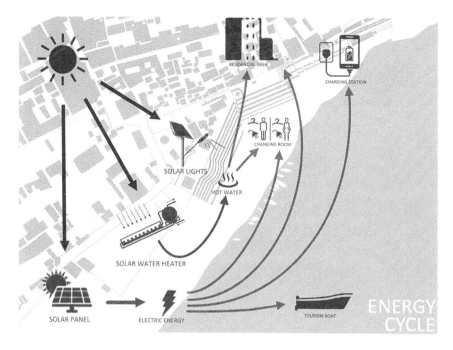

Figure 14.2 Solar energy cycle. Source: Author (graphic drawn by Wanhui Zuo)

The solar lighting system will increase safety and accessibility as the *ghats* are presently unevenly lit. Bollard lighting on pilgrim trails, lighting, spot lighting on heritage buildings and signage, and evacuation lighting during emergencies, can be installed at low cost. Solar energy can also supplement the grid in buildings along the *ghats*. The *ghats* support many small vendors who do brisk business by selling tea and snacks. They rely on biomass fuels such as coal or wood for cooking which contribute to air pollution and is harmful to the health of particularly those cooking. Solar energy harnessed by sun ovens is clean and inexpensive and can be used for cooking in small restaurants and food stalls. Their parabolic shape, shiny surfaces, and adjustability allow harvesting the maximum amount of sunshine at any given time of the day. Sun ovens can also be mounted on mobile carts thus increasing the mobility of vendors.[9]

Flora

Vegetation is sacred in Hinduism because of its associations with powerful gods and goddesses. The *bel* tree is symbolic of Varanasi's ruling deity Shiva and numerous shrines containing *shiva lingas* are found under *bel* and other

Figure 14.3 Shrines under the *neem* tree at Assi Ghat. Source: Author

trees such as *pipal* and *neem* (Figure 14.3). The densely built-up southern part of the *ghats* has a few tree shrines; the tree cover becomes denser near the confluence of Varana and Ganga where seasonal farming occurs on the river embankment. The flood plain soils on the east bank support seasonal farming as well. Among flowers, lotus of the Nymphaeaceae family occupies a very special position in the Indic culture. Its emergence from the waters signifies the birth of cosmos itself. As the seat of gods and goddesses, particularly Vishnu's consort Lakshmi, it is a symbol of stability as well the dynamic life force, *shakti*, that caused form to emerge out of watery chaos (Sinha 2006, 40). Marigold flowers are extensively used in ritual worship. The symbolic meanings of plants rest upon their material attributes, in particular their curative and remediating powers; it is this aspect that can be harnessed for addressing air and water pollution on the *ghats*. The utility of vegetation can extend beyond its shade giving property and as a source of food (cereals, vegetables, fruits) to its ability to filter pollutants and use treated waste as fertilizer.

Untreated sewage from the city is the major source of pollution; among others two major drains, Nagwa and Khirkhi, located at the southern and northern ends of the *ghats*, empty untreated sewage into the river.

In addition, domestic wastewater from the old city is carried to the river through surface drains (*nalis*). A large amount of solid waste is generated by the adjoining urban neighbourhoods. The shoreline at the popular *ghats* is littered with rubbish and it is not uncommon to stumble into rotting piles of ritual offerings. Local point source pollution stems from flower offerings, food scraps, plastic bags, clay pots, lumbar from old boats, washing of clothes and bodies, animal waste, ash from cremations, charred human flesh, and animal carcasses. Non-point-source pollution from industrial effluents and agriculture adds to the overall contamination levels, causing the biological oxygen demand (BOD) to increase by more than 500% in the Ganga after she passes through Varanasi.

Phytotechnology, i.e. plant-based system for remediation, is natural, uses solar energy, and is low-cost (Kennen and Kirkwood 2015, 7). Free surface water constructed wetlands are effective in wastewater treatment, easy to operate, and can be maintained by local communities. Warangal, a city in the southern state of Telangana, for example, effectively uses wetland vegetation to filter out contaminants in the wastewater (Jayakumar and Dandigi 2002). Wetlands can complement the industrial sewage treatment plants that are resource intensive in terms of energy use, space, and engineering skills. For example, where Assi Nala empties into the Ganga, phytoremediation will be useful in treating wastewater and sullage carried by the stream. The old streambed of Assi Nala that was channelized to build Ravidas Ghat can be restored and widened to construct rain gardens. This will enhance the water retention ability of Assi Nala and prevent flooding during monsoons. Plants in the rain gardens will absorb pollutants and filter waste.

At Manikarnika Ghat, popularly known as the 'burning ghat' because of its perpetual cremation pyres, the air is always thick with smoke and cinder from the burning logs, and the surrounding buildings are covered with soot with temperatures reaching 60 degrees Celsius in the scorching June sun. The dominant wind direction is from west to east, spreading smoke, ashes, and dust to the Ganga. Shade trees such as *neem*, *pipal*, and *banyan* will attenuate air pollution and improve the microclimate. Small gardens can be planted at *ghat* edges, especially where silt deposits after the monsoons and is difficult to remove. Local organic waste such as leaves, flower and food offerings, and clay pots that are strewn on the steps can be collected and treated in compost tumblers to fertilize the silt gardens here and at other *ghats*. The proposed compost program for transforming waste into a productive growing medium requires trash receptacles, compost bins, and tumblers to be located at strategic points along the *ghat* stretch.

Fauna

Bovines are an integral part of the *ghat* scene, ambling up and down the steps, resting on landings, eating out of garbage, munching marigold offerings, and bathing in the Ganga. Their ubiquity in public spaces and lack of restriction on their movements is based upon the respect accorded to the cow in Hindu culture. The cow is revered as mother or *gau-mata* and is given protection, her slaughter a grave sin in Hinduism. She is a symbol of earth in its generosity and support of all living beings, and of wealth and prosperity. She is associated with Krishna and ascribed divinity in her form as *Kamadhenu*, the cow of plenty who gives all asked for. Her milk and milk products are used in many sacred rituals and her dung is used as fuel, fertilizer, and disinfectant. Nandi bull, the vehicle of Shiva, is visibly present in sculptured stone at the threshold of shrines and temples on the *ghats*. Decorated cows are often paraded around and the chant *gau hatya pap hai* (killing cows is a sin) can be frequently heard at the conclusion of a ritual ceremony on the *ghats*.

The 'zoogeography' or study of cattle movement (Hui 2015) on the five *ghats* revealed dung piles in lanes and stepped landings, adding to other biodegradable forms of waste. The roaming cattle and goats at the cremation sites on Manikarnika Ghat disturb the solemnity of funerals, and at the other two crowded *ghats* – Dashashwamedh and Panchganga – add to congestion. At Assi and Raj Ghats, buffaloes are found in large numbers, resting and bathing in the river. The widespread presence of cattle means that their dung is produced in large enough quantity for it to be treated as a resource (Figure 14.4). Vandana Shiva describes the cow as a keystone

Figure 14.4 Cow on Manikarnika Ghat. Source: Author

species for sustainable agro-ecosystems and performing a critical function by converting organic matter into a nutrient for plants. The symbolic status of the sacred cow in Hindu society is derived from her utility as an economic resource.[10]

At the *ghats*, animal dung, instead of flowing into the Ganga, can be systematically collected and used to generate biogas and fertilize gardens. At Assi Ghat, a milk cooperative society is proposed in the adjoining *maidan*. Here the roaming buffaloes can be gathered in a buffalo farm, fed flower offerings from rituals occurring all over the *ghats*, and their dung collected in the biodigester. Anaerobic digestion will treat the manure to produce biogas that can be used for cooking milk-based sweets, in great demand as *prasad* (offerings) to the gods in Varanasi temples and shrines. At Adi-Keshav Ghat, a *gau-shala* (cattle shed) is part of a proposed ashram next to the Adi-Keshav Temple where *panchkroshi* pilgrims stop before completing the final leg of their circumambulatory journey around Varanasi. Cows will be an integral part of the *ashram* life, their milk, and curd and ghee made from it used in ritual ceremonies. Their dung as well as other organic waste collected from Raj Ghat will be treated to produce biogas and the digestate (residue left from the process of anaerobic digestion of biodegradable material) will be used as fertilizer in the community gardens planted on the embankment to grow fresh vegetables for the *ashram*.

Conclusion

The sustenance of cultural heritage embodied in the *ghats* is tied to planning policies and design interventions that can be effectively implemented using local resources and energy. The complex ecosystem of the *ghats* had evolved from self-sustaining cycle of human activities integrated with natural rhythms; it is unbalanced and stressed today as its carrying capacity has been exceeded many times over. Although the *ghats* as a public space are a civic responsibility, the government has been ineffective in filling in the vacuum left by the loss of royal patronage in independent India. Ganga's pollution is a grave threat to her centrality in Hindu thought as a symbol of purity. Rituals tied with holy waters are cornerstone of cultural heritage of Ganga and the *ghats*. Their long-term sustenance will be ensured when the concepts of *swatcha* (clean) and *shuddha* (pure) Ganga are aligned as they had been in the past. The resilience of this cultural landscape has been proved time and again; however, unprecedented levels of pollution may cause irreparable damage to this legacy.

The design strategies proposed in this chapter will conserve cultural heritage and protect the riverfront from any further development that is not ecologically and culturally sustainable. Designing with water, sun, flora, and fauna in mind would create new cycles of energy production and consumption and of waste transformed and recycled, all embedded within the age-old cycle of spatial practices tied with natural rhythms. This approach is in keeping with the Geddes' framework of bio-centric planning in which urban settlements are a part of regional ecology and where community development is a function of working with traditional practices in harmony with natural processes. On the *ghats* in Varanasi where nature is worshipped based upon the powers of Ganga to create and sustain life harnessing the energy of sun, it is important that the transcendent status of the sun and the river and their earthly physical forms are not considered separate, and their symbolic and utilitarian values be combined in planning for a healthy and resilient landscape.

Notes

1 Rta Ritu: An exhibition on Cosmic Order and Cycle of Seasons. Indira Gandhi National Centre for the Arts, New Delhi, January 4–March 30, 1996.
2 BBC News, July 1, 2014, Sudhiti Naskar, "The river where swimming lesson can be health hazard," accessed June 20, 2020, http://www.bbc.com/news/magazine-28112403
3 According to *The Times of India* (December 13, 2016) report by Binay Singh the environmental scientist B .D. Tripathi had conducted a study in 2009 of ambient air quality in the seven km long stretch between Assi and Raj Ghat and found the quantity of nitrogen dioxide to be 145-165 microgram per cubic meter and sulphur dioxide to be 135-155 microgram per cubic meter, far exceeding the permissible limit (80 microgram per cubic meter). Accessed April 25, 2018, http://timesofindia.indiatimes.com/city/varanasi/Varanasi-Allahabad-air-most-polluted-Study/articleshow/55952226.cms
4 Ganga has a coliform bacteria – indicating human and animal waste – count of more than 1.5 million per 100 ml. 500 coliform per 100 ml is considered safe for bathing. Biological oxygen demand (BOD) levels in the river are over 40 mg/l in Varanasi. They should be 2 mg/l or less.
5 The Ganga Action Plan was launched by the Prime Minister Rajiv Gandhi in 1986 for controlling pollution caused by industrial waste and domestic sewage from the many cities and towns on its riverbanks. The plan achieved limited success and the Ganga remains one of the most polluted rivers in the world.
6 HRIDAY comes under the Ministry of Urban Development while PRASAD is implemented by the Ministry of Tourism. Fourteen cities across India have been selected for their rich cultural heritage and domestic tourism, and 12 cities come under both schemes. HRIDAY aims for planning, development, and implementation of heritage sensitive infrastructure, especially sanitation services, in the core areas of the historic city. Accessed June 20, 2020, https://smartnet.niua.org/content/7663306a-7610-46a0-919f-d0688d9493e8. PRASAD is for rejuvenation and spiritual augmentation of important religious destinations by promoting sustainable pilgrimage and tourism. Accessed June 20, 2020, https://pradhanmantri-yogana.in/prasad-scheme/
7 www.dictionary.com
8 The project reports Ghats of Varanasi on the Ganga in India: The Cultural Landscape Reclaimed (2014) and Envisioning a Resilient Cultural Landscape: Ghats on the Ganga, Varanasi, India (2016) summarizing design studio work at University of Illinois at Urbana Champaign, USA can be accessed at: https://amitasinhaprofile.com/creative-projects/

9 This project was developed by Reina Patel in the Grand Challenge Learning Course for Freshmen (GCL 129) taught by Amita Sinha at University of Illinois at Urbana Champaign in Spring 2016.
10 Shiva, Vandana, 2002. "In Praise of Cowdung." ZNet Daily Commentaries, November, accessed April 25, 2018, https://zcomm.org/zmag/

References

Alley, Kelly. 1998. "Idioms of Degeneracy: Assessing Ganga's Purity and Pollution." In *Purifying the Earthly Body of God*, edited by Lance Nelson, 297–330. Albany, NY: State University of New York Press.

Alley, Kelly. 2016. "Rejuvenating Ganga: Challenges and Opportunities in Institutions, Technologies, and Governance." *Tekton*, 3: 108–123.

Conaway, Cameron. 2015. "The Ganges River is Dying Under the Weight of Modern India." *Newsweek*, September 23. Accessed June 20, 2020. http://www.newsweek.com/2015/10/02/ganges-river-dying-under-weight-modern-india-375347.html.

Eck, Diana. 1982. *Banaras, City of Light*. New York: Alfred Knopf.

Eck, Diana. 2015. "Ganga: The Goddess Ganga in Hindu Sacred Geography." In *An Anthology of Writings on the Ganga: Goddess and River in History, Culture, and Society*, edited by Assa Doron, Richard Barz, and Barbara Nelson, 233–251. New Delhi: Oxford University Press.

Hui, Rebecca. 2015. "MOOving Along: Following Cows in Changing Indian Cities." *Tekton*, 2(1): 08–24.

Jayakumar, K.V., and M.N. Dandigi. 2002. "A Study on the Use of Constructed Wetlands for Treatment of Municipal Wastewater During Summer and Rainy Seasons in a Semi-Arid City in India." *Global Solutions for Urban Drainage*, 1–13. Accessed April 25, 2018, https://doi.org/10.1061/40644(2002)27

Kennen, Kate, and Niall Kirkwood. 2015. *Phyto: Principles and Resources for Site Remediation and Landscape Design*. London: Routledge.

Khan, Naveeda. 2011. "Geddes in India: Town Planning, Plant Sentience, and Cooperative Evolution." *Environment and Planning D: Society and Space*, 29: 840–856.

Lannoy, Richard. 1999. *Benaras Seen from Within*. Seattle: University of Washington Press.

Ramakrishnan, P.S. 2003. "The Sacred Ganga River-based Cultural Landscape." *Museum International*, 55(2): 7–16.

Singh, R.L. 1955. *Banaras: A Study in Urban Geography*. Banaras: Nand Kishore & Bros.

Singh, Rana P.B. 1994. "Varanasi Cosmic Order and Cityscape: Sun Images and Shrines." *Architecture+Design*, November–December: 75–79.

Singh, Rana P.B., and Pravin Rana. 2006. *Banaras Region: A Spiritual and Cultural Guide*. Banaras: Indica Books.

Sinha, Amita. 2006. *Landscapes in India: Forms and Meanings*. Boulder: University Press of Colorado.

Sinha, Amita. 2015. "The Enacted Landscape of Varanasi Ghats: Beyond the Picturesque." Visual Arts Journal, India Habitat Centre, Special Issue on Art in Public Places, 12: 40–49.

Sinha, Amita. 2017. "Introduction." In *Cultural Landscapes of South Asia: Studies in Heritage Conservation and Management*, edited by Kapila Silva and Amita Sinha, 1–10. London: Routledge.

Tyrwhitt, Jacqueline, ed. 1947. *Patrick Geddes in India*. London: Percy Lund Humphries & Co, Ltd.

15 Conservation of Indo-Islamicate water experience

James L. Wescoat Jr.

Introduction

Water has fascinating, multifaceted roles in the heritage of the Indian sub-continent (Jain-Neubauer 2016; Wescoat 2013). Few pre-modern water systems continue to operate, but they still capture the imagination in ways that lead one to aspire to hear old fountains play again, see pools filled to the brim, and restore ancient wells that lifted cool groundwater for diverse human wants. Across South Asia, there are efforts to restore historical water systems to working conditions. In parallel with these sentiments are the concerns of conservation architects to waterproof structures in ways that minimize the damaging effects of water – wicking, rot, and mold (Wescoat 2007). Water simultaneously has life-fulfilling and destructive powers. Even damaging events like floods have long-term ecological and cultural significance. To move toward an integrated approach toward water herit-age management, it is useful to begin with a broad conceptual framework for water-conserving design. The first part of this chapter reviews the major components of a conceptual model of water-conserving design to set the stage for the emphasis here on a conservation case study of Indo-Islamicate water experience. The term 'Indo-Islamicate', coined by Marshall Hodgson (1975/2009, vol. 3, p. 87), refers to cultural engagement and exchange between Muslim and non-Muslim cultures in South Asia, like the Mughal-Rajput case study in this chapter.

Conceptual framework of water-conserving design

The conceptual framework developed here began with a presentation to the Indian National Trust for Art and Cultural Heritage (INTACH) that included a Venn diagram of three overlapping circles that connected water-works conservation with water resources conservation and water symbol-ism (Wescoat 2007). These represented three major interests in Mughal gardens and waterworks research during the late 20th century. As research advanced, the model expanded to encompass socio-economic livelihoods based on water, and phenomenological aspects of water experience (Wescoat

2014a). The culture of water as understood by cultural anthropologists and geographers encompasses all of these components, and it is conserved when all facets are integrated with one another. The role of conservation design lies in the center of the diagram, which is the locus of integration. It is useful to review briefly each of these heritage conservation subfields as a way of setting the stage for this chapter's emphasis on the relatively neglected topic of water experience.

Waterworks conservation focuses on physical structures such as fountains, pools, and channels that defined the layout and logic of extant gardens and historic urban landscapes. Monumental masonry aqueducts, *qanats*, and stepwells are fascinating structures in their own right, as well as functional means for supplying urban water needs. They are regarded as forms of water architecture as well as water engineering (Hegewald 2002). Even the infrastructure of springs, dug wells, water conduits, and drains can capture the human imagination. Conserving these water structures depends primarily upon the materials involved. Brick, stone, terracotta, lead, mortars, plasters, and so on – each has its own conservation methods and techniques (Wescoat 2013). The modern public aspiration is to make these structures hold and display water again. But as noted above, water also constitutes a threat to heritage conservation. Water seepage contributes to weathering, rot, mold, leaching, spalling, and related damages (Wescoat 2007). Conservation architects devote extraordinary efforts to waterproofing roofs, foundations, walls, and openings (Rose 2005). Should conservators seal a structure or let it breathe – bring water into historic structures or keep it out – maintain high indoor air pressure to reduce humidity, or draw humid air through a building to emulate historical processes of ventilation? Societies have developed many forms of water architecture that require a continuous balancing of these productive, protective, and destructive forces.

Water resource conservation: While the conservation of waterworks is important at historic sites, water resource conservation is important in all landscapes. Water is scarce worldwide, particularly in arid and semiarid regions like western India. The large amount of water that was used to irrigate historical gardens, fill cascades of tanks and fountains, and wash away waste through open drains may no longer be available or deemed to be reasonable uses of scarce and vulnerable water resources in modern times. Consumptive evapotranspiration by plantings and waterbodies are major losses from a local water system. Groundwater withdrawn in excess of recharge is unsustainable. Although often small by current standards of use, historical consumption must compete with current needs. In a related vein, historical water systems may be too small to make a significant contribution to modern water needs. Historical waterworks may no longer be physically strong enough to collect, store, and convey water supplies without substantial leakage and breakage, which leads some conservation architects to lay new pipe networks over old ones, or to cover them with a new layer of masonry that alters the dimensions and perception of the original water features. Important exceptions to these practices include the restoration of traditional rainwater harvesting systems that link surface collection with subsurface storage (Agarwal and Narain 1997).

Water budget analysis is a useful method for managing supply and demand in historical water systems (Wescoat 2000, 2014a). It uses simple mass balance computations to equate water inputs with outputs and changes in storage over time: I = O + dS. Water inputs to a site include rainfall, runoff from upper slopes, open channel flow, closed pipe delivery, and groundwater withdrawals. Inflows and outflows from tanks and cisterns are relatively easy to measure, though that is rarely done at historical sites. Estimating water outputs from evaporation, transpiration, and groundwater discharge is more challenging, but they too have well-developed hydrological methods. Soil moisture is often the most complex and variable factor to measure, even on small sites (Wescoat 2013). Recently, water quality and pollution have received more attention in historical waterways in India (Haberman 2006; Wescoat 2016, 2019).

Conservation of water livelihoods is perhaps the least developed and most socially important component of the model. It shifts attention from waterscapes of heritage consumption to those of heritage production. Production includes the work of human and animal labor lifting water from wells, channeling it to sites, irrigating plantings, and treating wastewater (Wescoat 2014b). Mughal paintings include portraits of *bhishtis* (water carriers), some of whom still sprinkle dusty streets and gardens from water-filled animal skins today. The fruits of their labors are largely accrued to owners and rulers. Out in the countryside, irrigators retained a fraction of their produce which was generally higher than that of dryland cultivators (Habib 1996).

Conservation of water experience is the focus of this chapter, so it will be briefly introduced here and elaborated in the next section. All modern accounts of historical water systems employ prose descriptions to spark the reader's imagination. Their power varies with the writer's abilities, taste, and contemporary standards. They often evoke monsoon skies and flowing, sometimes flooding, rivers, glistening surfaces, and relief from summer heat and dust. Experience links all other components of the conceptual model, but justice has not been done to it to date.

Conservation of water cultures and meanings

This encompassing category has received some research attention among architectural historians, for example in *Qur'anic* references to the rivers of paradise and their symbolic representation in gardens that await the faithful upon the resurrection (Blair and Bloom 2009). Yasser Tabbaa (1986) wrote an evocative essay on meditative qualities of water flows from the *shadirwan* and *salsabil* structures that carry water from the outlet in a wall, down a cascade of water pools and channels, which he associated with a religious philosophy of Islamic occasionalism. This interpretation suggested close connections between sensory experience and symbolic meanings, which are explored in greater detail below (Wescoat 2009).

Conservation of water experience

This section surveys approaches to water experience that have taken many forms in the diverse regions of Indo-Islamicate India, and it uses a case study to illustrate conservation implications. This culture area spans from Afghanistan in the west to Bengal in the east, and from Kashmir in the north to Deccan in the south, with all the languages and peoples these regions encompass. The term Indo-Islamicate embraces both the diverse Muslim religious traditions and their myriad interactions with Hindu, Jain, Christian, and local belief systems that have shaped the rich combinations of tangible, intangible, and living heritage in South Asia. It has been transformed but not erased by colonial and post-colonial modes of globalization. The case study considered here, the Nagaur Fort palace complex, has sultanate, Mughal, Rajput, and now modern layers of water heritage in the Marwar region of Rajasthan (Diamond 2008; Jain, Jain and Arya 2009; and Jasol 2018). It exemplifies the manifold character of Indo-Islamicate water systems and experience.

No single model of water experience can do justice to these diverse places, times, and traditions. Instead, this study offers multiple perspectives from, and on, Indo-Islamicate water experience at places that have Rajput and Mughal heritage, which can help interpret the diverse and dynamic qualities of water in ways that have relevance for contemporary water conservation practice in India. One of the challenging questions in case studies is how to initiate and organize the narrative. Should one start with historical evidence about when the complex was first built, and add new perspectives as it developed over time? Or should one begin with a hydraulic approach, i.e., with the sources of water supply, followed by their collection, storage, conveyance, and use? Or with water features and their experience as encountered by a visitor? Should one start with the experience of patron, worker, or visitor?

Each of these approaches has merits and challenges. This chapter starts with an account of the study area as it might be experienced by a modern heritage conservationist, traveling from the nearby metropolis of Jodhpur to Nagaur Fort and its various interior garden spaces. This may be regarded as a pragmatic or more precisely pragmatist approach because it starts with contemporary experience and the practical aim of conservation. The chapter then follows an historical trajectory, beginning with an early sultanate perspective on the citadel and its medieval tank complexes, followed by later Mughal and Rajput perspectives on Nagaur's palace garden water systems, as represented in paintings as well as the built environment, and concluding with the performative aesthetics of water experience and conservation practice.

The pragmatist approach

The pragmatist philosophy of experience focuses on dynamic interactions among society, environment, and everyday activities of making and doing (Dewey 1934/1958). It regards experience as a continuous process of inquiring, feeling, acting, and understanding. Some types of experience are problem-driven while others occur through an openness to spontaneous wonder. In both cases, the flows of human experience are analogous to flows of water experience. The pragmatist perspective on experience is well-aligned with the view of heritage conservation as a form of reflective problem-driven practice. While pragmatism has some of its roots in American philosophy, it has been influential for Indian intellectuals like Dr. Bhimrao Ambedkar who studied with John Dewey at Columbia University, and for the professions of architecture and planning (Ockman 2001; Stroud 2018). Philosopher Richard Shusterman (2003) has written about the links between pragmatist and classical Indian aesthetics. If the pragmatic method begins with a current situation and an initial, problematic, experience, it does not stop there. It searches for historical and geographical evidence (e.g., texts, paintings, field research, etc.) to deepen experience, clarify issues, and generate conservation alternatives.

This case study of Nagaur Fort began when modern conservationists were invited by the heritage site's owner, the Mehrangarh Museum Trust based in Jodhpur, Rajasthan. The road from Jodhpur northeast to Nagaur travels along the margin of a desert that receives only about 300 mm of rainfall per year. It has long been a pastoral landscape of goat, camels, and cattle herding. These rural lifeways are arduous, with occasional relief from monsoon rains, dug wells, and neem trees planted along the roadside. Many wells are brackish and their water undrinkable. After a short period of relief, modern tubewell irrigation and urbanization have depleted regional groundwater in ways that are poorly understood but profoundly unsustainable.

Conservation implications: It is important for heritage conservationists to compare knowledge and experience of historical water users such as herders and rural peasants with those of modern cultivators, townspeople, and tourists. Such comparisons have been made in the rainwater harvesting field (Agarwal and Narain 1997). For the practice of heritage conservation, it is likewise important to compare modes of sensory experience and traditional water knowledge with modern scientific measurement (e.g., for rainfall, evapotranspiration, groundwater, etc.).

Sultanate tank complex experience

Arriving at the city walls and gates, one crosses an important threshold between the desert and densely settled neighborhoods of Nagaur city that date back at least to the Sultanate era (11th–15th c.) (Shokoohy and Shokoohy 1993) (Figure 15.1). There are seven major extant earthen water tanks, impoundments created by damming shallow depressions in bedrock surrounding the city. Each tank complex has a major mosque, shrine, market, residential neighborhood, and special guilds of artisans nearby (Jain, Jain and Arya 2009). Some have stone *ghats* leading down to the water surface. In addition to their direct use and experience for

Royal tent area
Bakht Singh Mahal
Main water gardens
Entrance gate
Road from Jodhpur

Figure 15.1 Satellite image of Nagaur citadel, city, and tanks. Source: Google Earth, 2017.

drinking, bathing, and washing, they recharged shallow aquifers that were tapped by dug wells and manual water-lifting devices. These surviving waterbodies and their urban fabric and architectural surroundings parallel sultanate waterworks developments in the Gujarat and Delhi regions, notably those of Sultan Firoz Shah Tughluq in Delhi (reigned 1351-1388 CE).

Conservation implications: Large water bodies marked strategic junctions between urban centers and their arid to semiarid landscape environments. They were experienced as working socio-religious waterscapes. Today, these urban lakes are threatened by filling, encroachment, and pollution that makes some of them repelling. Tapping distant water supplies like the Indira Gandhi Canal that now serves Nagaur may accelerate this neglect of local waterworks. At the same time, rediscovery of historical rainwater harvesting systems has expanded rapidly over the past two decades with some encouraging conservation results (Agarwal and Narain 1997). Nearby mosques, shrines, temples, and community guilds have contributed to the historical experience of urban tanks, and restoring those relationships is a major opportunity and need for the quality of life in cities like Nagaur today.

Mughal water conservation experience

It has been vital to conserve these tanks throughout history, as evidenced in Mughal paintings and texts such as the *Akbarnama* (chronicles of Akbar, the third Mughal ruler [reigned 1556–1605 CE]). Upon arriving at Nagaur Fort, Akbar ordered that a tank outside the citadel gate be desilted (Abu'l Fazl 2011, vol. II, pp. 517–519). A double folio painting depicts that event and thereby sheds light on water conservation experience (Victoria and Albert Museum, 2020). On the left-hand sheet, the painting portrays the king on horseback supervising the work in an arid landscape where tents are pitched in the background. The king and his entourage face a set of workers and city residents standing outside the citadel walls, on the right-hand folio. The city enclosure is densely filled with stone masonry buildings and trees. The foreground of the painting focuses on mature men doing the desilting. They have strong figures, hoeing and hauling silt from the tank with energy. Some look to Akbar for direction and approval, but most know what to do and are doing it. The middle ground of the painting is dominated by Akbar and his retinue on the left and by townspeople on the right. This is the register of the benefactor and beneficiaries. In other episodes of this sort, Akbar participated in the work, as he enjoyed physical labor and crafts. The townspeople indicate their gratitude with outstretched arms. They include an old woman, possibly a widow in white, and a young woman. In the background, men come and go from the Mughal tented encampment and the citadel walls. This is a rare depiction of royal involvement in landscape maintenance, i.e., conservation, as compared with construction.

Conservation implications: This historical event and painting are instructive for Nagaur today. Its seven lakes have important heritage values, but they are in poor physical condition, as they were in Mughal times, suffering from experience-degrading siltation, algal blooms, and waste strewn around the banks. In historical terms it is important to learn whether and how historical conservation experiences of the type depicted in the *Akbarnama* paintings have been replicated. The argument here is that historical images can help inform and inspire water heritage conservation today (for examples, see Aga Khan Trust for Culture 2017; Baig and Mehrotra 2017; Singh 2016; and Wescoat 1989).

Rajput water conservation experience and aesthetics

Approaching the large entry gate to the Nagaur Fort citadel, one crosses a moat that captures surface drainage and recharges a shallow aquifer around the fort. Just within the gateway complex is a second rainwater harvesting area that captures runoff from areas surrounding the palace complex. Inside the palace, all of the rooftops drained into tanks and cisterns in the buildings and courtyards. One does not immediately perceive these runoff collection areas and methods, but the Mehrangarh Museum Trust which manages the site has analyzed and restored many of them (Jain, Jain, and Arya 2009). Walking uphill to the palace one passes through drought adapted-vegetation established by the Trust and its planting designer Pradip Krishen. These xeric plantings evoke images of desert plants outside the city walls in Mughal paintings, as compared with the lush forest and orchard plantings in Rajput garden paintings.

One enters a parade ground where modern tourist tents are pitched. Although this scene is partly a digression, it is worth noting that the luxury tents actually contribute to the conservation of water experience. Each tent is supplied with a bucket of hot water and another of cold water for bathing, washing, and minimal flushing into a temporary pit latrine. Occupants of these luxury tents thus consume about the same amount of water as modest households in the city of Nagaur today, as compared with the much greater amounts consumed in conventional heritage hotels.

Visitors then pass through forecourts that lead to private interior quarters of the Rajput ruler, family, and retainers, which the Trust and conservation architect Minakshi Jain (2009) have sensitively repaired. The first interior courtyard is associated with a palace building known as the Bakht Singh Mahal named after a long-ruling king (infamous for having killed his father). It has a medium-sized pool with nine fountains. Beyond that lies a small Shish Mahal *chahar bagh* garden, which provides valuable insights into the hybrid nature of Mughal-Rajput design, for it successfully combines a nearly symmetrical set of cross-axial water channels within an angular Rajput garden space (Tillotson 1987) (Figure 15.2). The conservation planting in this small Shish Mahal quadrangle by landscape and garden conservationist Professor

Figure 15.2 Conservation of the Shish Mahal garden and waterworks by Mehrangarh Museum Trust (author photo).

Priyaleen Singh (2016) echoes the hybrid aesthetic of Mughal-Rajput gardens by combining a few lush exotics like banana with a broader palette of drought-tolerant ornamentals like pomegranate and oleander.

Adjacent to the Bakht Singh Mahal courtyard is the main sequence of palace waterworks and gardens. This series begins with a large rectangular bathing tank in the foreground, which fills during the monsoon with rooftop runoff. It is followed by a *chahar bagh*-type space that has two quadrants of lotus pools and two quadrants of plantings in the middle ground, which is to my knowledge a unique innovation in Indo-Islamicate garden waterworks – half the garden is a sheet of water, and the other half a planted oasis. The sequence concludes on a raised terrace with a *baradari* (literally, twelve-doored) pavilion, and a deep square water tank with an island platform in the background. These generous bathing pools distinguished Rajput palaces from their Mughal counterparts, which were experienced more by viewing than bathing. Mughal and Rajput paintings often convey lavish water systems that cascade through multiple garden terraces. Monsoon rain still flows off the rooftops through carved drainage spouts into the bathing pools below. As with the lushly imagined vegetation in Rajput paintings, monsoon water paintings freeze instants in time that may have occurred once a year, if that often, for the viewer's sustained enjoyment. Rajput paintings at Nagaur thus conveyed idealized forms of water experience.

In comparison with Mughal paintings of waterworks that were largely meant to be seen, Rajput tank scenes involved swimming and boating scenes of the raja, harem, and nobility (Figure 15.3). It would be easy to interpret these courtly water scenes as simple images of luxury. However, recent research on Rajput painting and aesthetic theory suggests that they were more than that in experiential and cultural heritage terms.

Interpreting Rajput Water Experience: At one level, Rajput bathing scenes may be taken at face value as hedonistic scenes of royal excess, meant almost exclusively for royal consumption. Some of these images have a discordant quality, as for example when water-abundant paintings were produced during a period of drought and suffering (see the insightful interpretation of Indic water imagery during Little Ice Age weather anomalies by Sugata Ray [2017]). However, an important exhibition of Marwar paintings, titled *Gardens and Cosmos: The Royal Paintings of Jodhpur*, interpreted water imagery on the following six levels (Diamond 2008):

- Royal pastimes in the gardens at Nagaur Palace – these are scenes of water play among the raja, consorts, and nobles. They include bathing; splashing; colored *holi* water fights; and boating among water lilies, fish, ducks, and egrets. They have strong erotic qualities.
- Gardens for divine play – these paintings resemble court scenes but feature gods like Krishna in positions otherwise occupied by the raja, lending support to ideas of divine kingship. What distinguishes these scenes from royal pastimes is that they often include separated lovers – Radha

Figure 15.3 Maharaja Bakht Singh rejoices during Holi. 1748–1750. Courtesy: Mehrangarh Museum Trust.

and Krishna, Krishna and the *gopi*s, Rama and Sita – and thus signify the experience of intense longing. Water channels separating lovers in these contexts are expressions of that desire, and of potential release and fecundity.

- Maharaja Man Singh and the *Nath*s – these paintings indicate a shift from pan-Indian gods toward regional *Nath* guruship in early 19th-century Marwar, in which *Nath* gurus were represented in divine waterscapes.
- Origins of the cosmos – include paintings that substitute a *Nath* guru in increasingly abstract cosmological paintings of the origin of the universe, for example, floating in the cosmic sea with a lotus emanating from his navel.
- Mapping the cosmos – paintings depict miniature buildings and urban scenes mapped onto *chakra*s of the human body, and mandalas that flow through the subtle channels (*nadi*s, lit. streams or flows) of the body, linking microcosm to macrocosm.
- Sacred sites and cosmic ocean – these paintings are the most abstract representations of water and space. They include simple panels of gold or silver paint, marked only by wavelike line patterns. These cosmic ocean paintings signify the waters of primal existence, far beyond those of earthly gardens and territories, which are experienced as imaginable but not perceptible in the built environment.

Conservation implications: This progression of water representations connects courtly pleasures with divine waterscapes and abstract ideals. It suggests a chain of water signification that challenges conservationists to contemplate higher levels of symbolic water experience. One can readily imagine conservation strategies to fulfill the first level of courtly water experience, and by extension gardens of divine play that evoked *bhakti* experience of devotion and longing. Each of the higher levels of abstraction requires deeper contemplative inquiry that lies beyond mainstream professional conservation practice. To bridge these gaps, Diamond (2013) suggests in a separate study the links between aesthetic experience and yoga practice. Some of these higher levels of experience were historically transitory. Nath gurus and their followers were controversial and ultimately eclipsed in their own times. Conserving these transitory and transgressive modes of water experience is challenging.

Varieties of Indo-Islamicate water experience

In addition to the different levels of water experience discussed above, from royal to cosmological, there is a wide variety of Rajput traditions some of them closely engaged with the Mughal court, as in Marwar, while others purposely distanced themselves from Mughals and Sultanate courts, like the Mewar court in Udaipur (Aitken 2010). The latter refused political and marriage alliances with the Mughals, and their painting traditions maintained a strong sense of difference that adds complexity to the representation and challenges of conserving water experience. While some rulers and courts had hybrid sensibilities, others leaned toward experiences associated with one sect or another. These sects and associated styles often crossed over between categories of Hindu, Muslim, Jain, and Nath.

This crossover is exemplified by the passionate engagement of Hindu courts, including Mewar, with the Muslim love story of Layla and Majnun (Aitken 2010). Like divine romances of endless longing, the story of Layla and Majnun progresses from friendship to devotion, madness, death, and only at that point is unity attained among self, other, and the divine. Most relevant for this study is the consistent water imagery in paintings of Layla and Majnun, which frequently depict a stream flowing from a raised throne-like mound where the lovers were seated, often within a wider dryland setting. Many of these images have a superabundance of birds and animals, reminding the viewer of Majnun's ability to communicate with animals (like King Solomon). Streams in these images have a sacred quality, either by virtue of being a source of purity, issuing from the earth, or as a source of plenitude for the teeming life within the painting. Sultanate and Hindu courts in the western and southern Deccan regions of India likewise distinguished themselves from one another, to a greater or lesser degree, and their water imagery can be interpreted through similar comparisons of hybridity and distinction.

Conserving the performative aesthetics of water experience

At least one more level of water experience was present at Nagaur. Images of courtly and divine love evoke a fundamental dimension of human experience that is introduced but not fully developed above, which has had limited recognition in heritage conservation to date. The foundations of aesthetic philosophy in much of South Asia rest on theories of *rasa*, literally meaning flavor or taste but theoretically translated as aesthetic emotion (Pollock 2016). Developed initially as a theory of poetry and performing arts circa 300 CE, *rasa*s give rise to nine stable emotions that include the erotic, violent, heroic, macabre, comic, tragic, fearful, and fantastic. They are further associated with 33 transitory emotions, and thousands of combinations of emotions in poetry and drama. These theories reached their apogee in medieval Kashmir in the 11th to 14th centuries CE, but they continued to influence the arts through the Mughal-Rajput period in works like the *River of Rasa*, which have relevance for this paper (Pollock 2009). It is important to acknowledge that these literary philosophies said little about painting and almost nothing about the aesthetics of architecture or the natural world (Pollock 2016). Early *rasa* theorists further argued that the experience of literary works has little to do with environmental experience, which is different. However, some modern commentators discern strong historical connections among literary, performing, and visual arts (Taylor 1997). The pragmatist approach also argues for inherent continuities between aesthetic and environmental experience (Ray 2017).

With this brief perspective on *rasa* theory in mind, one may return to the dual nature of erotic water experience represented in Nagaur paintings, i.e., as the love-making in courtly scenes, and the longing expressed in divine scenes. Water is found in both forms of aesthetic emotion, as reflected in images of turbulent monsoon skies, sheets of rain falling from the clouds, large rivers teaming with water creatures, and small water bodies that separate or unite the lovers in a painting. Each of these water elements conveys erotic *rasa* experience that entails a succession of stable and transitory emotions. By this line of reasoning, it may not be an accident that the courtly throne space in Figure 15.3 is rubbed out, leaving an empty abstract white void.

Conservation implications: These intangible heritage values have relevance for understanding the performative aspects of water in architecture and in landscape conservation. They also help us understand the emotional dimension of conserving water experience. And finally, they require an appreciation of the seasonality of water heritage. For years, I have asked colleagues whether Indo-Islamicate garden pools and fountains should be kept full or allowed to vary with the seasons, down to the point where they are completely dry. As a landscape architect I regard empty pools and channels as beautiful volumes of space that invite the imagination of water experience, past, present, and future. Empty water bodies are certainly more beautiful than those that are leaking and algae-clogged. While granting this last point, my interlocutors rarely acknowledge the beauty of empty water

Figure 15.4 Jal Mahal tank empty before the monsoon. Source: Author.

bodies. They insist on the pleasure and delight of full water tanks, channels, playing fountains, and lush plantings. Their arguments are supported by Mughal and Rajput paintings that never show a water tank empty or only partially full. However, my argument is supported by the more abstract Marwar paintings and countless Indo-Islamicate poems of erotic longing in seasons of dryness, yearning for rain and streams, and for love that is absent. The point to underscore here is that all of these varieties of aesthetic water emotions can be realized through the conservation of seasonal water experience, that is, through the anticipation and appreciation of experience of water in different amounts and qualities at different times of year and times of day. This takes a crude form in the sound and light shows at historic fountains and gardens. A more sophisticated conservation approach takes seasonal water fluctuations seriously, asking visitors to contemplate the reduced volumes and sprays in the month of May as akin to longing in the wider world, which are transformed by pulses of water during evening visits, and overflowing abundance during the monsoon rains (Figure 15.4).

Conclusions

This chapter has explored the cultural bases for conserving Indo-Islamicate water experience in India, and for moving beyond the reliance on ad hoc and subjective interpretations of how historic water systems might have been experienced in various places and times. It offers multiple perspectives on water experience, and multiple implications for conservation practice. The Nagaur Fort case study provided a good test of this approach by showing how different traditions have contributed to manifold dimensions of water experience. Here are the main conclusions for conservation practice in India:

First, it is helpful to begin with a modern pragmatist approach, which underscores the central role of experience in heritage conservation and problem-solving. This chapter provides evidence for its relevance in India as well as elsewhere.

Second, I have demonstrated the value of reconstructing historical layers of water experience. In the Nagaur example, this approach began with sultanate tanks along the city walls and proceeded to the Mughal, Rajput, and modern conservation of those tanks centuries later. The layers vary in different periods and places, but each place has its own logic and body of evidence that must be reconstructed to do justice to it in its own terms. The juncture of Mughal and Rajput waterworks design at Nagaur was compared with other Rajput artistic traditions, recognizing their differences as well as their potential relevance for one another in the broad expanse of Indo-Islamicate culture.

Third, constructing a spatial narrative of travel through historical palace-garden courtyards at Nagaur helps shed light on the complex interactions among water experiences in different places, times, and cultures. Even within a specific cultural tradition, water experience has ranged from erotic royal bathing scenes to increasingly abstract cosmological levels of water experience evidenced in paintings. Conservationists should be explicit about how they are translating experiences from one artistic medium, such as painting, to the built environment, and to the development of conservation alternatives. These translations can be checked and refined through peer review.

Most importantly, it is indeed possible and meaningful to draw conservation implications from historical ideas and evidence about water experience. These implications help us understand why many conservation architects, and the public at large, have regarded water as one of the most exciting and challenging aspects of the field (Jain-Neubauer 2016; Jain, Jain, and Arya 2009, 188–193). Indeed, the work of conservationists brings us back to the pragmatic, and principled, problem-driven approach to water heritage conservation that opened the chapter. The next generation will more fully realize how conservation practice can be coupled with and guided by the depth and breadth of Indo-Islamicate water experience.

Acknowledgements

I am grateful to the leadership of the Mehrangarh Museum Trust for its generous support in Jodhpur and Nagaur, to Elizabeth Moynihan for introducing me to the Trust, and to colleagues Debra Diamond, Minakshi Jain,

Karni Singh Jasol, Pradip Krishen, and Priyaleen Singh on conservation at Nagaur Fort, and to all of the participants in the Nagaur Garden Workshop, especially Milo Beach who first opened my eyes to the experience of Mughal and Rajput painting. Manish and Ashima suggested many helpful editorial improvements.

References

Abu'l Fazl. 2011 reprint. *The Akbarnama*, translated by H. Beveridge, 3 vols. Delhi: Low Price Publications.

Aga Khan Trust for Culture. 2017. *Humayun's Tomb Conservation. Rethinking Conservation*. New Delhi: Mapin Publishing.

Agarwal, Arun, and Sunita Narain. 1997. *Dying Wisdom: The Rise, Fall, and Potential of India's Traditional Water Harvesting Systems*. New Delhi: Centre for Science and Environment.

Aitken, Emma. 2010. *The Intelligence of Tradition in Rajput Court Painting*. New Haven: Yale University Press.

Baig, Amita, and Rahul Mehrotra. 2017. *Taj Mahal: Multiple Narratives*. Delhi: OBI.

Blair, Sheila, and Jonathan Bloom, eds. 2009. *Rivers of Paradise: Water in Islamic Art and Culture*. New Haven: Yale University Press.

Dewey, John. 1934/1958. *Art as Experience*. New York: G.B. Putnam.

Dewey, John. 1925/1958. *Experience and Nature*. New York: Dover.

Diamond, Debra. 2008. *Garden & Cosmos: The Royal Paintings of Jodhpur*. Washington: Smithsonian Institution.

Diamond, Debra. 2013. *Yoga: The Art of Transformation*. Washington: Smithsonian Books.

Haberman, David. 2006. *River of Love in an Age of Pollution*. Berkeley: University of California Press.

Habib, Irfan. 1996. Notes on the Economic and Social Aspects of Gardens in Mughal India. In *Mughal Gardens-Sources, Places, Representations, and Prospects*, edited by James L. Wescoat Jr., and Joachim Wolschke-Bulmahn, 127–138. Washington: Dumbarton Oaks.

Hegewald, Julia. 2002. *Water Architecture in South Asia*. Leiden: E.J. Brill.

Hodgson, Marshall G.S. 1975/2009. The Venture of Islam, Volume 3: *The Gunpowder Empires and Modern Times*. Chicago: University of Chicago.

Jain, Minakshi, Kulbhushan Jain, and Meghal Arya. 2009. *Architecture of a Royal Camp: The Retrieved Fort of Nagaur*. Ahmedabad: AADI Centre.

Jain-Neubauer, Jutta, ed. 2016. *Water Design: Environment and Histories*. Mumbai: Marg Publications.

Jasol, Karni. 2018. *Peacock in the Desert: The Royal Arts of Jodhpur, India*. Houston: Museum of Fine Arts.

Ockman, Joan, ed. 2001. *The Pragmatist Imagination: Thinking about Things in the Making*. New York: Princeton Architectural Press.

Pollock, Sheldon I., Trans. 2009. *Bouquet of Rasa and River of Rasa by Bhanudatta*. New York: New York University Press, Clay Sanskrit Library.

Pollock, Sheldon I. 2016. *A Rasa Reader: Classical Indian Aesthetics*. New York: Columbia University Press.

Ray, Sugata. 2017. Rupa and Rasa, Material Form and Theological Aesthetics: Picturing the Riverscape in the *Isarda Bhāgavata Purāṇa*, ca. 1560. In *Aesthetic Practices and Spatial Configurations: Historical and Transregional Perspectives*, edited by Hannah Baader, Martina Becker, and Niharika Dinkar, 145–158. Bielefeld: Verlag.

Rose, William. 2005. *Water in Buildings*. New York: Wiley.

Shokoohy, Mehrdad, and Natalie Shokoohy. 1993. *Nagaur: Sultanate and Early Mughal History and Architecture of the District of Nagaur, India*. London: Royal Asiatic Society.

Shusterman, Richard. 2003. Definition, Dramatization, and Rasa. *Journal of Aesthetics and Art Criticism* 61(3): 295–298.

Singh, Priyaleen. 2016. *Conserving Water System in Medieval Gardens*. Presentation at Aurangabad. https://www.youtube.com/watch?v=8g3x9NrGyUQ&t=4s. Accessed October 31, 2017.

Stroud, Scott R. 2018. Creative Democracy, Communication, and the Uncharted Sources of Bhimrao Ambedkar's Deweyan Pragmatism. *Education and Culture* 34(1): 61–80.

Tabbaa, Yasser. 1986. The Salsabil and Shadirwan in Medieval Islamic Courtyards. *Environmental Design: Journal of the Islamic Environmental Design Research Centre* 2: 34–37.

Taylor, Woodman. 1997. Picture Practice: Painting Programs, Manuscript Production, and Liturgical Performances at the Kotah Royal Palace. In *Kotah: Its Gods, Kings and Tigers*, edited by Stuart Cary Welch. New York: Asia Society, 61–72.

Tillotson, Giles H.R. 1987. *The Rajput Palaces*. New Haven: Yale University Press.

Victoria and Albert Museum. 2020. *Purification of Kukar Talao (Water Tank) at Nagar, Rajasthan in 1570 by Order of the Emperor Akbar*. Websites: http://collections.vam.ac.uk/item/O9595/painting-kesav-kalan/ and http://collections. http://vam.ac.uk/item/O9596/painting-kesav-kalan/. Accessed June 20, 2020.

Wescoat, James L. Jr. 1989. Picturing an Early Mughal Garden. *Asian Art* 2: 59–79.

Wescoat, James L. Jr. 2000. Waterworks and Landscape Design at the Mahtab Bagh. In *The Moonlight Garden: New Discoveries at the Taj Mahal*, edited by Elizabeth B. Moynihan, 59–78. Washington: Smithsonian Institution and University of Washington Press.

Wescoat, James L. Jr. 2007. *Conserving Mughal Garden Waterworks. Sir Bernard Feilden Lecture Publication*. New Delhi: Indian National Trust for Art and Cultural Heritage.

Wescoat, James L. Jr. 2009. Searching for Wisdom in Mughal-Rajput Waterworks: East-West Interdependencies. In *Bau- und Gartenkultur zwischen "Orient" und "Okzident": Fragen zu Herkunft, Identität und Legitimation*, edited by Joachim Ganzert and Joachim Wolschke-Bulmann, 187–213. Munich: Martin Meidenbauer Verlag.

Wescoat, James L. Jr. 2013. Water and Waterworks in Garden Archaeology. In *Sourcebook for Garden Archaeology Methods, Techniques, Interpretations and Field Examples*, edited by Aicha Malek, 421–451. Bern: Peter Lang.

Wescoat, James L. Jr. 2014a. Water-Conserving Design: Contributions of Water Budget Analysis in Arid and Semi-Arid Regions. In *Out of Water: Design Solutions for Urban Regions*, edited by A. Chaouni, and L. Margolis, 163–173. Basel: Birkhauser Press.

Wescoat, James L. Jr. 2014b. *Water and Work in the Mughal Landscape. Professor M. Athar Ali Memorial Lecture.* Aligarh: Aligarh Muslim University, January 21.

Wescoat, James L. Jr. 2016. Barapula Nallah & Its Tributaries: Watershed Architecture in Sultanate and Mughal Delhi. In *Marg Special Issue on Water Design: Environment and Histories*, edited by Jutta Jain-Neubauer, 84–95.

Wescoat, James L. Jr. 2019. Nallah to Nadi, Stream to Sewer to Stream: Waterscape Planning in India and the United States. In *Water Histories: The Materiality of Liquescence*, edited by Sugata Ray, and Venugopal Maddipati, 135–158. London: Routledge.

16 Restoring and nurturing the 'nature–human' bond through conservation of historic gardens

Priyaleen Singh

Introduction

Human relationship with nature has found expression in various civilizations articulating the cultural aspects of the people, the place, and the time. The garden was one such expression that wove its way into urban, suburban, and rural landscapes of India imparting them with many secular and sacred meanings. These historic gardens and landscapes today form a very important part of our valuable heritage. However, they are also perhaps the most vulnerable and the first to suffer destruction. While it takes effort, time, and resources to demolish a heritage structure for new development, a historic garden, seen by most merely as an open space, offers itself readily for new development. It is also true that in the recent past there has been a lot of concern expressed towards the preservation and restoration of historic buildings; this has unfortunately not been echoed in the case of historic gardens. Gardens have been generally treated as appendages to and surroundings of historic sites and not as important entities in their own right.

Gardens present special challenges for conservation. Neglect of historic gardens due to lack of finances or inadequate understanding of their heritage significance has led to the loss of many an historic garden. Because of improper perspectives on urban planning, leading to land speculation and unplanned building activity, historic gardens continue to be destroyed. Subdivision of large garden estates resulting in fragmentation and multiple ownerships, as well as changes in land use within and around them, are all factors responsible for this loss. Often, after centuries of existence outside or on the fringes of city boundaries, historic gardens end up in an urban context with large populations. While it is sometimes possible to retrieve and restore a landscape that has suffered neglect, further insensitive urbanization invariably tends to destroy not just the visual but also its ecological and physical contexts. Compared to other kinds of tangible heritage, historic gardens, owing to their living character and need for constant renewal, are more susceptible to change. While natural change, essentially seasonal, involving change in colour, form, and textures is one of the attractions in a garden, misguided interventions at conservation result in the gardens losing their original meaning, content, and form. There is an urgent need to recognize historic gardens as an important part of our heritage so that all the issues mentioned above can be appropriately addressed.

In wanting to restore historic gardens to their former glory, the rationale cannot simply be in valuing them as symbols of the past. While acknowledging authenticity as integral to good conservation practice, wherein any restoration activity calls for pedantic accuracy, historic gardens also have to be integrated with contemporary living in meaningful ways. Such an approach would ensure that they remain relevant in present-day life as perhaps they once were in the past. In doing so, the rationale for their conservation has to build upon the strengths of meanings and messages of the past, especially those messages that are universal and timeless and thus of significance today, and connect them to the new associations and meanings the garden would have acquired over time. In respecting these messages, conservation will also ensure that the historic gardens retain their spirit and authenticity. Taking these essential aspects of landscape preservation into account, this chapter discusses four garden projects. 'Conservation of gardens of Taj Mahal' in Agra was a collaborative research project between the Taj Mahal Conservation Collaborative (TMCC) and the Archaeological Survey of India (ASI), undertaken in 2001, with the aim to understand the spirit of a Mughal garden. The larger objective was to frame a methodology of research for Mughal gardens in the Indian subcontinent, which would assist with their recognition and restoration. This template has since successfully been applied to the recently restored gardens of Itimad-ud-Daulah's tomb garden in Agra. The next two examples pertain to the conservation of Moolsagar in Jaisalmer and Chokhelao in Jodhpur, both historic gardens managed by the Mehrangarh Museum Trust (MMT), a private trust dedicated to the cause of conservation. The projects were initiated and executed over a span of seven years from 2002 to 2009 and exemplified a model of heritage economics, wherein garden conservation became an economic asset for the custodians of heritage. The fourth example of conservation of Ram Bagh in Amritsar, a public garden under the joint management of three public institutions, Municipal Corporation Amritsar, the ASI, and the Punjab Heritage Tourism Board (PHTB), was undertaken in 2008 as part of the initiative by the PHTB to conserve the heritage of Punjab for its people. As a major public space in the city of Amritsar, its conservation demonstrated how the interests of heritage and local communities can converge seamlessly, resulting in an improved quality of life for all users of these gardens.

Gardens of Taj Mahal, Agra

The gardens of Taj Mahal, one of the most famous monuments of the world, have undergone transformations of the kind perhaps seen in most Mughal gardens in the subcontinent. As part of the project to prepare a conservation plan for the gardens, it was important to understand the various forms of symbolism all Mughal gardens are replete with. But what was equally necessary was to understand the idea and the vision of the designer in planning the Taj and its gardens. Research undertaken as an essential part of the project included sourcing archival material, which included memoirs of Mughal emperors and accounts by court historians; accounts of European travellers contemporaneous to the period of construction as well as later visits to the gardens; reference to gardens in literature and poetry; paintings of the Mughal and Colonial period depicting the Taj and other Mughal gardens; drawings of the late Mughal and Colonial period showing changes made to the gardens; correspondence during the Colonial period between individuals and institutions responsible for looking after the Taj and photographs from the early twentieth century (Singh 2002). Archival analysis supplemented by on-site studies revealed that the full view of the Taj Mahal was meant to be enjoyed from only two locations. A comprehensive visual and spatial analysis revealed that the first view was from the threshold and platform at the south entrance gateway. Another, relatively unknown but the most important view was from the north, from the Mehtab Bagh across the river. From other points from within the *char bagh*, the planting ensured that the Taj Mahal would have been visible in parts with the minarets rising above the trees. In continuation with the narrative of paradise, perhaps this was a deliberate ploy to create the effect of the mausoleum to be poised above ground, the minarets symbolically connecting to the heavens and making it appear as a light, ethereal structure, despite its monumental scale. Flowers and fragrance were also essential features of the paradise gardens. And collectively, they encapsulate the spirit of the Mughal garden. This involved all the five human senses in their experience: touch that caressed the textures of various plants; taste that savoured the fruit from the orchards; sound that heard the rustling leaves in the wind, the chirping birds and the music of the bubbling fountains; smell that intoxicated the spirit through the fragrance of the scented flowers; and sight that appreciated the riot of colours as the flowers bloomed. The entire experience within the gardens of Taj Mahal, however, changed in the late nineteenth century in colonial India because of the introduction of new vocabularies of design that were essentially derived from the western cultural context. Under the ASI, an institution set up by the British in 1861 CE, the early twentieth century saw 'thinning out' of vegetation so as 'not to impede the view of the tomb', resulting in the groves being replaced by manicured lawns and neat seasonal beds as borders, aligning with the then prevalent English tastes in landscape design. This was done despite the knowledge that in the Indian climate a stretch of lawn is, in ecological terms, an unsound proposition compared to a grove of trees, as it makes high demands on the maintenance and water resources and provides no benefits, including shade and edible flowers and fruits. Schemes continually made in 1906, 1914, and 1923 planned these borders, rockeries, and trellises within the newly introduced lawns (Archaeological Survey of India 1871–1940).

The changes in the ambience were not just confined to the visual. The preferences of the British also contrasted sharply with the olfactory tastes of the Indians. Artist William Hodges, on a visit from England, in painting the views of the Taj Mahal was 'repelled' by the overpowering perfume within (Desmond 1992), which presumably came from the *juhis, chamelis* and marigolds, species integral to all the pre- Mughal and Mughal gardens. As a result, the gardens of the Taj Mahal lost out on all the scents and fragrances that would have significantly contributed to the sensuous experience of the site. What was lost was also a garden that had been originally designed to encourage a participatory relationship with nature, wherein the landscape had engaged all the human senses.

In conceptualizing a conservation plan for the gardens of Taj Mahal, while authenticity of experience remained an important consideration, the rationale guiding its return to the paradisiacal experience of Mughal times was based on their continuing appreciation by society in providing a richer and wholesome human experience. In making a case for the restoration of the Mughal gardens to the administrators and users, the argument was based on very pragmatic and easy to comprehend reasons. The restoration scheme based on exhaustive archival research reasoned that the groves of trees made more ecological sense in their demand for and consumption of water and in supporting a richer and more diverse ecosystem, than an expanse of lawns (Figure 16.1). These trees would additionally provide much-needed shade to the thousands of visitors in tropical climate. And most importantly, the plantings would also provide increased biomass around the Taj Mahal, as regulated by an order of the Supreme Court of India to counter the air pollution that is turning the white marble of the mausoleum yellow.[1] The orchards would also help sustain the gardens as they originally did over three centuries ago. And in bringing alive the gardens, the visitors would have another dimension of the site to appreciate, that of the gardens. This would help take pressure off the mausoleum which appears to be the only attraction today. It were these arguments that proved that in terms of landscape design the Mughal garden experience is still of relevance today and therefore worth conserving.

Moolsagar, Jaisalmer

The Mughal *char bagh* influenced various other contemporaneous garden styles, all of which however developed in very different cultural milieus. One such case was the Rajput garden design tradition, of which Moolsagar and Chokhelao are two examples. Moolsagar, in the desert of Jaisalmer, is a Rajput garden first laid out in 1780 CE by Maharawal Moolraj II, the thirty-fourth ruler of the Bhati dynasty of Jaisalmer. The garden exists as a series of four introverted courtyards all planned within the framework of a *char bagh*, with exquisitely carved stone pavilions and water features. While most of the features belong to the eighteenth and nineteenth century,

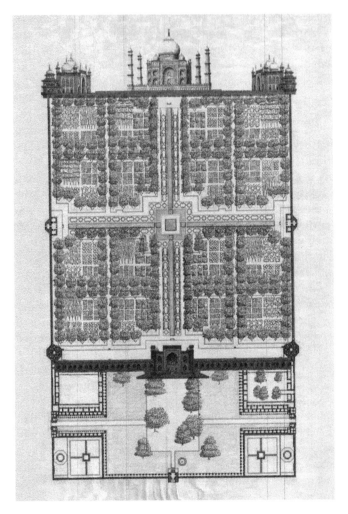

Figure 16.1 Proposed planting plan for the *char bagh* at Taj Mahal. Source: Author

elements such as the cast-iron pergolas added in the early twentieth century represent the colonial phase. The garden was originally designed as a royal retreat, visited frequently by the royal family of Jaisalmer. In recent times, in the absence of that patronage, the garden had been neglected and erased from the memory of the local community. Sited in the middle of the desert, it had over time ceased to have any connection with the surrounding habitation. Interestingly, the rationale for its conservation was derived from this very isolated existence, as an oasis in the middle of hostile desert environs. The restoration of the garden, which in itself would not have perhaps been as meaningful an exercise, was given additional value by treating it as a heritage property central to an ambitious tourism project. The project brief of simply restoring the garden was reformulated to conserve the garden as a setting for a camp resort, with the intention of demonstrating how a historic garden could help guide a very sustainable model of heritage tourism.

The innovative idea of a resort in tents, with its near total 'reversibility', represented one of the least intrusive of all adaptive reuse interventions and thus was considered most appropriate from the conservation point of view (Figure 16.2). Over thirty luxury handcrafted tents were planned around the historic garden, set in orchards of *jamun* and citrus trees, evoking the era of royal tent camps set within and around the gardens, seen both in the Mughal and Rajput periods. They were strongly reminiscent of the royal durbars held under colourful awnings in garden settings. In their very limited use of water, the accommodation in tents was also an example of eco-friendly

Figure 16.2 Moolsagar: tents pitched around the *char bagh*. Source: Author

tourism. In respecting the local ecology, the attempt was to recover the original *jhalara* or stepped well and the *baoli* within, which had been silted to the extent of being completely buried and lost under the sand. They were desilted, cleaned, and brought back as the principal source of water to the garden, as per the original scheme of over two hundred years ago.

The restored landscape today is striking in its setting as a camp resort. The first courtyard, with the *chandni* and hibiscus shrubs set against the backdrop of banana trees, take on the form of a *mehtab bagh*, perfect for a moonlight experience with its aromatic blossoms of *kamini* and *mehndi*. The second courtyard, as the day garden, provides colour through extensive use of flowers identified from Rajput paintings and laid out in the *char bagh* style. The third courtyard respects its colonial layer and retains the memory of time with vivid colours of the bougainvillea rambling over the cast iron pergola. The fourth courtyard is kept as the outdoor dining space. The *jhalara* has also become an important part of the site as a venue for evening cultural performances. The landscape revival has thus woven into the fabric of a historic garden the local traditions, festivals, and cuisine, all of which celebrate the local Marwari culture. The project has given a new lease of life to the historic garden, illustrating how sensitive interventions, proper management, and responsible design can make it possible to both respect the past, and address contemporary visitor needs without compromising on the principles of conservation. The garden today has become an economic asset for the owners and the model is being replicated in other historic gardens in Nagaur and Chokhelao.

Chokhelao, Mehrangarh Fort, Jodhpur

Located at the foot of the Mehrangarh Fort in Jodhpur, Chokhelao is a garden originally laid out as a *char bagh* in 1739 CE by Maharaja Abhai Singh. It was successively modified until 1844 CE. Subsequently, over a century of neglect had completely destroyed all the planting, retaining only some of the features in the garden such as the pathways and terraces. Situated in an arid region, water in the garden was celebrated by locating the well in the centre, even as the garden lacked features such as running water or large water bodies. As a result, vegetation was the prime element contributing to the experience within the garden. Research of the planting systems and patterns, done primarily through analysis of paintings, poetry, and literature as well as royal records of the court, revealed that the garden was a celebration of nature creating various moods through the use of plant material which had a range of qualities such as sacred, medicinal, culinary, cosmetic, and aromatic. This illustrated an approach which was aimed at maintaining a holistic and balanced human existence in harmony with nature. It was this message that had to be continued and which provided the rationale for the conservation of the garden. The garden had been a backdrop to the exuberance of life as it unfolded through activities such as music and dance in pavilions, romantic interludes in the shade of trees, celebrating festivals, and holding *durbars* on platforms under colourful awnings. The proposal, referring to paintings of the period, choreographed the planting in each of the sunken planting beds to achieve an attractive mix of colour and texture, fragrance and utility (Crill 2000). The lower terrace was laid out as a *mehtab-bagh* or moonlit garden, with banana trees edging the *kamini* and *chandni* infill, to be enjoyed during the night. The upper terrace, as a day garden had a riot of colours and trees and fruit orchards for shade (Figure 16.3).

The restored garden, while evoking the memory of a bygone culture and tradition also attempted to illustrate how aesthetics of a landscape are not

Figure 16.3 Chokhelao: Upper terrace before and after restoration. Source: Author

simply based on the visual narrative, but in involving all the human sensory perceptions make for a more complete and worthy experience. The garden has become an asset for the Mehrangarh Fort and an important destination in its own right. Over the last few years the garden has been one of the venues of the Rajasthan International Folk Festival (RIFF) organized by the Mehrangarh Museum Trust. As a compatible activity, the festival enhances the value of and experience within the garden. A restaurant has been functioning in the wing adjoining the Chokhelao Palace, in the uppermost terrace, while an exhibition on Rajput gardens which would interpret the history of both Chokhelao and other contemporaneous Rajput gardens in the region is envisioned in one of the pavilions set within the outer walls. The garden, hitherto neglected, is now a major destination in the tourist trail that winds through the Mehrangarh Fort.

Ram Bagh, Amritsar

Ram Bagh in Amritsar was planned and laid out in 1848 CE as the summer palace of Maharaja Ranjit Singh, who chose to have his capital in Lahore. It was a *char bagh* with fruit trees planted in the quartered planting beds functioning as orchards.[2] However, in the early nineteenth century, in the colonial period, when the garden came to be known as the Company Bagh, the spatial character of the garden was completely altered with the insertion of winding shaded avenues with extensive planting alongside, replacing the *char bagh* (Figure 16.4).

Ram Bagh is a garden of historical significance as it is one of the few surviving gardens in the region from the nineteenth century Sikh state under Maharaja Ranjit Singh. It has an identity as an indigenous adaptation of the Mughal *char bagh* style, qualifying it as an important example of the Sikh provincial garden style of the nineteenth century, the template of which was used in the following years in gardens at Patiala. The garden is of ecological significance as it exhibits an unusually large collection of trees planted over a hundred years ago that today qualify as rare and fine specimens in terms of their size and age. Owing to the presence of these large trees, the garden today hosts an ecosystem and biodiversity in flora and fauna, rarely seen in densely built-up urban areas like the one surrounding the garden. The garden has cultural and utilitarian significance as it is used extensively by the community today. It also has economic significance with its rich collection of fruit trees. In 2008, it existed as a generic open green space, since its historic form and design had been compromised due to several recent insensitive interventions by the Amritsar Municipal Corporation.

The spatiality of the *char bagh* had further been altered drastically with the introduction of new roads fragmenting the garden into smaller parcels by the agency, compromising the legibility of the site as a historic *char bagh*. As a result of this alien road network, the historic main gate of the garden had been reduced to a traffic island. The changes also included introduction

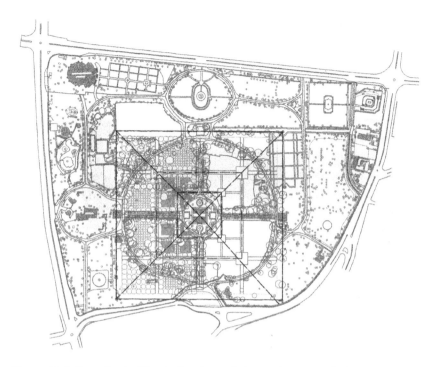

Figure 16.4 Ram Bagh: Plan indicating existing and proposed planting to restore the spatial character in two quadrants of the *char bagh*. Source: Author

of several incompatible land uses with their associated structures like the clubs, a skating rink, cages for a now abandoned zoo, and children's play areas. The boundary walls of the historic *char bagh* of Maharaja Ranjit Singh's period had also completely disappeared and the garden boundaries were redefined by a wall with a completely non-contextual design. The historic water channels were filled in, and the axiality, which was an integral part of the *char bagh*, had been altered beyond recognition. New water features such as tanks and fountains not in keeping with the historic character appear to have been randomly introduced at several places within the boundaries of the historic garden.

As part of the conservation plan, the spatiality of the *char bagh* had to be restored by removing the present circulation system and retrieving the pathways based on archaeological evidence that formed the cross axis of the *char bagh*. The tennis courts within the historic *char bagh* area had to be relocated outside in the buffer zone. The structures of the abandoned zoo had to be removed and its land was integrated as part of the larger buffer zone created beyond the historic *char bagh*. The boundaries of the *char bagh* also had to be accentuated in order to restore the spatiality of the historic design. Archaeological excavations were carried out at select points to reveal the cross axial water channels, which were vital in understanding the spatial layout of the site. The few surviving historic fountains were studied so as to be able to restore all the other fountains. Plant species, which were in consonance with those planted in the historic *char bagh* of the nineteenth century, were planted. The garden, once a suburban entity, is today located in a major commercial area. This location was seized as an opportunity to draw in a range of visitors into the historic garden as part of the heritage outreach programme. Two parking areas were located in the buffer zone in a discreet and visually screened area that offered a veiled view of the gardens from the commercial street.

Ram Bagh is an important open space for the residents of Amritsar. With the conservation of the *char bagh*, it was considered essential to control entries into the garden complex as a means of maintaining them. In formulating strategies for new ticketed entries, it was felt that movement from the surrounding areas would diminish significantly. In order to respect and honour locals' use of the site and association with it through time, it was decided to enforce ticketing for specific hours, to ensure the visitors coming to view Maharaja Ranjit Singh Garden got the necessary ambience to appreciate it, while at other times, the local residents could use it freely without much visitor traffic. A critical part of the conservation strategy was respecting the multiple layering of historic periods without undermining the spatial integrity of the site. Some of the memory of the colonial period circulation system was preserved by retaining all the trees planted during this period, designating them as heritage trees and integrating them with the proposed planting plan, so as not to disturb the ecology of the site.

Conclusion

At the outset it might appear that tourism underpinned all the examples discussed. While tourism is indeed a very seductive proposition in handling heritage, it need not and should not be the only framework for conservation of historic gardens. In conserving these gardens, while tourism had a role to play, it was the universal and local messages that emphasized and enhanced the conservation strategy. The universal messages are those that impact humanity as a whole and concern ecological design, the participatory relationship of people with nature, and traditional knowledge systems and their continuing relevance in contemporary times – all of which are amply illustrated in historic gardens. The local messages, on the other hand, are those that dwell upon the very specific and immediate cultural contexts of the gardens. For example, in the case of Ram Bagh in Amritsar, the garden is a major recreational open space for the local people, and in Chokhelao in Jodhpur, the garden is the only green space for the residents and visitors to the Mehrangarh Fort.

Since these gardens are continuously being embedded in new contexts and facing new challenges, their conservation has to be spelled out against the larger backdrop of urban development, where conservation becomes relevant in economic, social, and most importantly in ecological terms. Design with nature has clearly come a long way from being a 'subject' replete with social meanings and intents, to a superficial 'object' today, packaged for sale and conspicuous consumption. Conservation of historic gardens provides us with an opportunity to retrieve the richness of relationships with nature that existed in the past and to recover the local natural and cultural ecologies. In the narrative of changing perceptions towards nature in the city, historic gardens have immense potential as sites where natural and cultural heritage can be effortlessly woven into the everyday lives of the people and integrated with the contemporary open space systems as outdoor living spaces. These gardens, if conserved properly, would continue to function as the lungs of the city, demonstrating the value of open space and urban farming for an improved urban ecology.

In pitching for the conservation of historic gardens – a few of which continue to provide the much-needed open spaces in the urban concrete jungle – it is important to recognize their relevance as markers of history and cultural continuity. Historic gardens are a measure of a better quality of life, one that has not yet lost contact with elementary human needs and perceptions. And the success of conservation of historic gardens lies in keeping alive this spirit, one that can still touch the soul of urban living and existence.

Notes

1 For details see Supreme Court order passed on May 5, 1993 in response to Writ petition (civil) no. 13381 filed in 1984 by M.C. Mehta.
2 See *Gulgashat-i-Punjabc.1858* Acc. No. 38 in Maharaja Ranjit Singh Museum, Amritsar

References

Archaeological Survey of India Reports. 1871–1940. ASI Archives, New Delhi.
Crill, Rosemary. 2000. *Marwar painting, A history of the Jodhpur style.* Mumbai: India Book House Limited.
Desmond, Ray. 1992. *The European discovery of the Indian flora.* New York: Oxford University Press.
Singh, Priyaleen. 2002. *Gardens of Taj Mahal, concept plan* for TMCC. Unpublished report.

17 Community led sustainable cultural tourism

Ananya Bhattacharya

Introduction

Tourism has emerged as a leading socio-economic sector for creating jobs and regenerating rural and urban areas. The United Nation's World Tourism Organization (UNWTO) estimated over 1.5 billion international tourists worldwide in 2019 (UNWTO 2020). The agency also reported that tourism is contributing toward ten percent of global GDP and seven percent of global trade, and one in ten jobs (UNWTO 2017, 3). As a worldwide export category, tourism ranks third after chemicals and fuels, and ahead of automotive products and food. In many developing countries, tourism is in the top export category. Tourism represents 40% of service exports for emerging economies while the global average is 30% (UNWTO 2017, 6). Tourism is the primary source of foreign earnings in the world's 49 least developed countries and the principal export in over eighty percent of developing countries. UNWTO suggests that tourism will play a major role in poverty alleviation owing to its relative importance in contribution to the GDP in developing countries. Tourism is labour-intensive and creates jobs for the poor. Tourism development is also responsive to assets like biodiversity, cultural heritage or landscape and generally benefits sectors like agriculture, handicraft, transport and others. These factors, along with the size and growth of the sector, makes tourism a viable option for economic development in poor communities (UNWTO 2010, xiv).

The agenda for sustainable development, called Sustainable Development Goals (SDGs) for 2030 and set by the United Nations, also recognizes this important link between culture, tourism, and development. Cultural heritage and tourism are mentioned in the SDGs 8, 11, 12, and 14. Sustainable tourism and promoting local culture and products are recognized in Target 8.8 for Goal 8 which is for inclusive economic growth and employment opportunities. Goal 11, stipulating building inclusive and sustainable settlements, sets Target 11.4 to strengthen efforts to safeguard cultural and natural heritage. Target 12.b of Goal 12, deals with sustainable consumption and production patterns, and mandates developing and implementing tools to monitor sustainable development impacts for sustainable tourism in creating jobs and promoting local culture and products. Goal 14, dealing with conservation and sustainable use of the oceans, seas and marine resources, mentions sustainable management of tourism in target 14.7 (UN 2016).

A growing number of tourists are travelling to experience other cultures and are inclined to contribute to local economies through their travel. According to a survey of 69 member countries by UNWTO, 47% of international arrivals are considered to be "cultural tourists", that is travellers who participate in a cultural visit or activity as part of their stay. The overall growth in tourism is much larger for those countries that specifically feature cultural tourism in their marketing policy (66%) than for other countries (17%) (UNWTO 2018, 22).

The Association of Southeast Asian Nations (ASEAN) Community-Based Tourism Standard defines Community-Based Tourism (CBT) as "tourism activity, community-owned and operated, and managed or coordinated at the community level, that contributes to the well-being of communities through supporting sustainable livelihoods and protecting valued socio-cultural traditions and natural and cultural heritage resources". Natural heritage of a place includes landscapes, flora and fauna while cultural heritage is manifested in the built environment, relics, collections, festivals and practices, knowledge, and living experiences. UNWTO has adopted the following definition of cultural tourism:

> Cultural tourism is a type of tourism activity in which the visitor's essential motivation is to learn, discover, experience, and consume the tangible and intangible cultural attractions/products in a tourism destination. These attractions/products relate to a set of distinctive material, intellectual, spiritual, and emotional features of a society that encompasses arts and architecture, historical and cultural heritage, culinary heritage, literature, music, creative industries and the living cultures with their lifestyles, value systems, beliefs, and traditions.
>
> (UNWTO 2018, 16)

Growth of cultural tourism involves promoting awareness and supporting conservation of cultural heritage as there is continuous innovation and creation of new tourism products in response to the demand for new experiences by the tourists (UNWTO 2012, 1). The residents share with the visitors their history, folklore, art, craft, living habitats, cuisine, and other cultural aspects. The community earns from tourism products and the hospitality services they offer. The community as a whole benefits from an equitable share of tourist income though individual roles vary from business owner to employee. Community-based tourism brings business opportunities by creating the need for hospitality and transport services. The process generates shared resources for local economic development. The recognition and respect of local heritage not only strengthens community identity and pride but also mobilizes their action as key stakeholders in conservation and protection. Local leadership and ownership, and concern for natural and cultural ecosystems, might lead to development of sustainable infrastructure, and transparent decision-making supporting integration of aspirations and concerns of the community in larger developmental plans.

This chapter draws from the experiences of a social enterprise, banglanatak dot com,[1] working across India for fostering inclusive development using culture-based approaches. Its flagship initiative "Art for Life" (AFL) is revitalizing intangible cultural heritage and developing community-based creative enterprises, and has led to the emergence of artist villages as cultural destinations. The following sections outline approaches for developing cultural tourism by engaging local community and supporting growth of cultural enterprise based on local cultural assets to ensure creation and equitable distribution of tourism wealth and promotion of responsible tourism.

Synergizing cultural and economic development

Intangible cultural heritage reflects a community's wisdom, skills, and spirit. It is manifested in the oral traditions, performances and visual arts, traditional craftsmanship, knowledge of nature and universe, sports, rituals, and festivals. India is a land of living heritage. Rural and indigenous communities across India are bearers of a rich heritage of oral traditions, folk songs and dances, traditional crafts, and knowledge of nature. Oral traditions, songs, and dances are associated with life cycle events like birth, marriage, death, nature and traditional knowledge of farming, weaving, and other livelihoods. Unfortunately, with fast-paced lifestyle and urbanization, the folk art traditions are becoming moribund. With change in values and lifestyles, festivals, and ceremonies, there is loss of both traditional contexts and patrons. The situation is no different in the eastern Indian state of West Bengal, which is bordered by the countries of Nepal, Bhutan, and Bangladesh and the Indian states of Odisha, Jharkhand, Bihar, Sikkim, and Assam.

West Bengal, with a population of 91 million people (Census 2011), is rich in natural and cultural heritage and diversity. However, the tradition-bearers and practitioners are challenged in pursuing their arts and crafts while dealing with poverty and lack of opportunities to practice or perform. The Art for Life (AFL) initiative was launched in 2005 to address these challenges by safeguarding folk art traditions and synergizing cultural and economic development. The safeguarding and community development models used by the AFL initiatives work at three levels, namely: (a) revitalising the art traditions through strengthening skill transmission, documentation and promotion; (b) empowering the artists with skills and resources for innovation and creative enterprise development; and (c) transforming artists' villages into vibrant rural cultural hubs. While safeguarding intangible cultural heritage has changed the status of the individuals from daily wage earners to artists, promotion of artist villages as cultural destinations has led to the transition of the economically marginalized areas into local growth nodes in underserved areas. Development of community-led cultural tourism has assisted in rebranding the artist villages as cultural hubs.

Art for Life started with 3200 performing folk artists pursuing six folk traditions as part of an initiative to revive and revitalize performing folk traditions as a viable livelihood (banglanatak 2015). Between 2005 and 2008, the initiative was supported by the Eastern Zonal Cultural Centre, which is an autonomous body under the Ministry of Culture, Government of India. The European Union further supported the initiative between 2009 and 2011. In 2013, with the support of the Department of Micro, Small, and Medium Enterprises under the West Bengal State Government, and UNESCO, the Art for Life model was replicated in eleven clusters of traditional crafts for developing rural craft hubs (banglanatak 2014). Seeing the success of the rural craft hubs, the AFL model was further replicated to cover 15,000 practitioners of traditional art and craft in 15 districts of West Bengal for formation of rural craft and cultural hubs (banglanatak 2017). Additionally, the model was replicated in the state of Bihar, involving approximately 1500 artists. Art for Life was recognized as a good practice in the non-farm sector by the World Bank supported Bihar Innovation Forum-II (World Bank 2014). The Tourism Department of the Government of Rajasthan has partnered with UNESCO New Delhi to replicate the model in four districts of western Rajasthan since 2019 (*Business Standard* 2019).

The following stories share the unique journeys of the artist communities living in the villages of Naya, Charida, and Gorbhanga that emerged as heritage destinations frequented by students, artists, musicians, photographers, art lovers, and tourists. The artist villages are sustainably run by the community through performances or craft-making, and festivals celebrating various art forms. Naya is the village of a community of scroll painters and storytellers called Patuas. In Charida, the artistes specialize in making ornate masks used by Chau dancers. (Chau is a vigorous and acrobatic dance form originating in martial arts and popular among the indigenous people of Chota Nagpur Plateau in central India). Gorbhanga, a quaint village on the India-Bangladesh border, is a hub of Fakirs who are minstrels whose Baul and Fakiri songs promote divine attainment through love for humanity (banglanatak 2016). Chau dance and Baul and Fakiri music are inscribed in the UNESCO Representative List of Heritage of Humanity (UNESCO n.d.). The three villages showcase a unique model of developing cultural tourism based on intangible heritage (Figure 17.1).

Artist villages as destinations

Village of Patuas at Naya

There are 60 families of storytellers and scroll painters living in Naya village in the Pingla block of Paschim Medinipur district. Their tradition is called Patachitra. *Pata* means cloth and *Chitra* means painting. The painters

Figure 17.1 Village locations. Source: banglanatak dot com

belong to a community of Patuas with the surname Chitrakar (meaning "painter"). The village is small with five rows of painted mud and brick houses. The Patuas of the Medinipur region of West Bengal are followers of Islam. However, they traditionally paint stories from Hindu mythology on long scrolls and sing them as they unfurl the scroll. The songs are known as *Pater Gaan* (*Gaan* meaning "song"). Until a few decades ago, the storytellers used to go from door to door and sing songs in exchange for paddy and alms.

With the advent of modern entertainment and electronic media, the storytelling tradition lost its audience. Most of the Patuas took up vocations like pulling carts, hawking, or daily wage labour. By 2005, there were only around 20 artists at Naya who pursued their traditional crafts of which only two knew the songs. Patachitras were sold to art collectors who visited the village and few handicraft fairs. The families were extremely poor and their average monthly income was less than INR 500. In 2005, the Patuas started relearning their art and tradition from veterans as part of the Art for Life initiative. They also revived the tradition of extracting natural colour from seeds, flowers, leaves, bark, and stones. They nurtured their skill of depicting stories using a series of painted frames and accompanying songs. Today, the Patuas not only paint mythological tales but also paint and sing on themes ranging from the COVID-19 pandemic (*The Indian Express* 2020), climate change to gender rights, and prevention of violence against women and children. They paint biographical tales of icons like Gandhi, Tagore, and Martin Luther King. Their paintings adorn festival venues and international airports. Textile, apparel, and accessories with Patachitra motifs have also become popular in diverse markets.

The number of painters in Naya now exceeds 200, and women are in a leading role. While their mothers and grandmothers used to assist the men in making colours and painting, the women of Naya have stepped out of their homes and now travel across the country and the world. Naya is today a vibrant Patachitra hub. The average income of a Patua family is now around INR 15000–INR 20000, while leading artists typically earn above INR 30000 per month. It may be noted that Naya does not have any natural or built heritage to boast of, but the visitors are interested in the creative experience offered by the storytellers. Regular visitors include art lovers, students of schools and colleges, art historians, and designers. They marvel at the scrolls, listen to the songs, and learn about making colour from leaves, flowers, and seeds. The name of the annual village festival is 'POTMaya', a term meaning the wonder of the scrolls in Bengali. Since 2010 the festival has been held during a weekend in November and has become very popular. Today the village has a two-storey folk art centre exhibiting old and new scrolls with boarding and lodging facilities. The Patua families also offer homestay to visitors (UNWTO 2012,49, Times Now 2012) (Figure 17.2).

Village of Chau mask makers at Charida

Charida village in Baghmundi block of Purulia district lies near the Ayodhya hills in the Chota Nagpur Plateau. Ayodhya, with its verdant forests, is a popular tourist destination. The hydroelectric power plant at Purulia provides a panoramic view of the area. The 115 families of mask makers live in rows of tiled brick houses along the state highway close to the hills. The art of Chau mask-making started in Charida around 150 years ago, during the rule of King Madan Mohan Singh Deo of Baghmundi. The masks portray gods, goddesses, demons, and also animals and birds. Chau dance originated as a martial art and is vigorous and acrobatic. Over time, due to social and economic reasons, Chau dancers lost the traditional patronage of landlords and local kings and there was less opportunity to perform and practice. In time, the nuances of the dance, where the indigenous tribes used to emulate the movements of birds and animals like peacocks, tigers, and monkeys, were lost.

Figure 17.2 Tourists at Naya village of Patuas. Source: banglanatak dot com

Since 2005, the veteran Chau dancers have led the story of change as part of the Art for Life initiative. They were successful in training the younger generation in songs and dances and storytelling. Today there are hundreds of Chau groups regularly performing locally and across the globe. They participate in pageants and have even been featured in Indian cinema. With the growing popularity of Chau dance, the mask makers also enjoy year-round demand for the masks. Tourists visiting Ayodhya Hills purchase their craft at the Chau Mask Hub of Charida that started hosting an annual Chau Mask festival in 2014. The folk art centres at the Bamnia and Maldi villages in the Purulia district host annual week-long Chau dance festivals in December. In spring, Purulia's landscape gets a radiant orange makeover from the "flame of the forest", the Palash flower. Currently several village festivals are organized in Purulia every year celebrating Holi with folk songs and dance. The district now offers new livelihood opportunities through tourism development integrating built, natural and cultural heritage. Tourists enjoy the beauty of the red plateau dotted with hills and the dense deciduous forests. They visit the ruins of Santhal shrines and relics of Jain and Buddhist temples (UNESCO 2011). The resorts also offer programs of vibrant folk songs and dances (Figure 17.3).

Figure 17.3 A mask shop at Charida. Source: banglanatak dot com

Village of Fakirs at Gorbhanga

Gorbhanga in Karimpur-I block of Nadia district is near the India–Bangladesh border. The area has lush green paddy fields and mango, banana, and date palm plantations. In the winter months, golden yellow mustard fields delight travellers. The village is home to around 60 Fakirs who sing *Marfati* songs, which preach the importance of surrendering to the *Murshid* or *Guru* for attaining the divine. The underlying philosophy is that if the *Shariyat* is the tree, then *Marfati* is the fruit. The urge to be one with God, the pain of not attaining Him and the appreciation for the *Guru* – are all conveyed through their music. Fakirs reject all divisions of caste, creed, and religion and believe in knowing God through profound admiration, self-sacrifice, and faith.

A decade ago, the *Fakirs* of Gorbhanga village were ostracized from the orthodox communities. Their children were not allowed in schools, and their lifestyle and traditions were questioned. Opportunities to sing and perform in traditional assemblies or *Akhras* dwindled over the years (most of the *Fakirs* who were pursuing their tradition were old). When the Bauls of Nadia were mobilised to work together to promote their tradition, the leading singers conducted training sessions over three to six months using the traditional *guru–shishya parampara* (master–disciple tradition). The new learners were trained not only in lyrics, tunes, and rhythms but also oriented to the philosophy and meanings conveyed by the songs. The songs, which were orally transmitted until then, were documented. Capacity building addressed acquiring skills in stage performance and studio recording. Baul and Fakiri singers have now become popular among the youth and also at international peace concerts and music festivals. Musicians from across the globe travel to the village to collaborate with these talented musicians. The locals of Gorbhanga, who once derided the *Fakirs*, today respect the acclaimed artists who have emerged as music icons from daily wage earners. The folk art centre at Gorbhanga has become a centre for music tourism where artists from across the world visit for cultural exchanges and collaborative productions (TEDx Talks 2015).

Village festival as branding strategy

Eighteen village festivals are now annually held in different artist villages and they are emerging as cultural destinations. Many of the art forms practised in these villages have been languishing because the traditions or rituals they are part of are dying. The newly planned festivals create fresh contexts for practice and performance. The village festivals have given a positive identity to the underdeveloped hamlets. Festivals and promotional events are not only about showcasing or sale of products but also about familiarizing the tradition-bearers and practitioners with the creative sector, and audience choices and preferences.

Organization of such village festivals and facilitation of participation of artistes from other parts of the country and abroad also create a high level of public awareness. All these festivals enjoy extensive local and national media coverage. They are featured in travel and airline magazines and social media, which has led to increased visits of domestic and international tourists. The festivals are held annually over a weekend in the same month every year. The communities manage all operational issues and logistics. Visitors can partake in one-to-two-hour workshops in various traditional arts or crafts featured in the festivals. For instance, in Naya there are workshops on making natural colours. The festivals also showcase arts and crafts from other regions and cultures and thus create a scope for interaction and cultural exchange. The process helps in developing an understanding of the dynamics of culture, place, and society in different environments. The tradition-bearers gain confidence and increased ability to understand their own cultural context and innovate. In the process, the villagers often become aware of other ethnicities and cultures for the first time.

The first time a festival was held in each village, it had a ripple effect in the neighbouring towns and villages. The neighbouring communities knew about the artist communities but paid little heed to their dying traditions. Hearing about the festivals, they ventured into the villages and were impressed by the artistry and repertoire of the products. Local recognition and appreciation are extremely important for ensuring sustainability. The experience of the Art for Life initiative shows that while a small percentage of artists go outside to perform, most enjoy income from local performances and craft sales. Websites and social media also support global connections. For example, upon learning about "Folk Holi" festival through social media, a group of young Malaysians came to the spring festival in Purulia (Tripsavvy n.d.). An Italian university has been sending student groups to learn about Patachitra since 2015 (Alma Mater Studiorum– Universitàdi Bologna 2018). Today's globalized world thus offers tremendous opportunity for attracting a wide range of visitors.

Creating cultural spaces

Setting up and promotion of folk art centres embodying community identity is an effective strategy for bundling intangible cultural heritage for tourism (UNWTO 2012). The folk art centres established in the villages provide spaces for practice, promotion, and performance of the arts and crafts. The centres have been set up on land donated by the community. The state government of West Bengal has supported development of these centres under a program of developing common facility centres for micro enterprises (banglanatak 2014). The folk art centres showcase local culture, support practice and promotion, and offer places of stay to visitors. They promote and support year-round tourism to the village. Artist collectives registered as not-for-profit societies have been formed to manage the grassroots-level micro enterprises, folk art centres and village festivals to support promotion, networking, and destination management. They liaise with the district authorities, relevant government departments, and local governance structures for funding and technical support. They manage delivery of cultural products and services which range from performances at events or coordinating tours as well as managing sales of crafts (Figure 17.4).

Mitigating potential risks

While there are numerous benefits as outlined earlier, heritage-based tourism products must also avoid the risks of de-contextualization, misrepresentation, or over-commercialization. There is always much concern about commercialization of living culture when developing cultural tourism. Tourism development often creates a narrative in which the community has no control (Clyde 2011). The artist communities must not be further marginalized through lack of choice for participation in tourism activity or in everyday life. Cultural commodification results in production or performance exclusively for tourists and as a result they lose the cultural and spiritual significance. There are always risks of adverse impact on environment sustainability owing to the excesses of mass tourism, as in reserves like Masai Mara (Onchwati 2010) or coral reefs in Thailand's popular KohKhai islands and some of the world heritage sites

Figure 17.4 Visitors at a folk art centre. Source: banglanatak dot com

like Machu Picchu (WTTC 2017,19). Many Asian cities have had to deal with gentrification as a result of tourism owing to escalating rent or cost of living (UNESCO 2016, Yanwei Han & Wei Zhong 2015). Power relations also impact partnership between the practitioners and bearers of heritage, and the tourism stakeholders. Issues of misappropriation of cultural assets for profit, loss of control on cultural identity and rights are the other challenges. Privacy and safety challenges for both visitors and residents also need to be addressed in developing village tourism (UNWTO 2012, 10–16).

Such risks have been mitigated to varying degrees by facilitating community participation and consultation for developing village tourism. When the village festivals were initiated, three-to-five-day workshops were held with the community leaders, village youth, and women to sensitize them to the potential impacts of increased visitations, ways to understand and manage expectations, and develop tourism products and itineraries. Young men and women who mostly had secondary level education were trained in spoken English in month-long workshops. Several workshops were held to train the villagers on hygiene and cleanliness, safe drinking water, sanitation, and waste disposal. Additional workshops were held to build communication skills. Youth were trained for interpretation of local cultural and natural heritage. Capacity building, in this context, has to imbibe acceptance of social, religious, linguistic differences and respect for the "other" or hitherto unknown cultures and traditions. Tourism education and skills training of local youth is important for maximizing local socio-economic benefit (UNWTO 2010, 54–56). An analysis of 136 award-winning tourism practices across continents show the majority focused on training local youths in hotel management, tour operation, guide services, and life skills. Others worked on strengthening local value chain and support small businesses to integrate in the supply chain (Levy and Hawkins 2010).

Partnership building is a critical component of developing community-led tourism. Tourism investments are usually concentrated in infrastructure and services, rather than in engaging communities or building partnerships. This hampers sustainability and achieving optimal outcomes. The disconnect between the administration and those in charge of governance of culture and tourism sectors poses another challenge. Sector collaboration may ensure investment of tourism revenue in management and conservation of cultural and natural heritage. Multi-stakeholder partnership is also necessary to build sustained systems for mentoring rural and indigenous people to start new businesses in creative industries, and cultural sectors like community museums, heritage education, design, multimedia, and hospitality. Partnerships are also needed for building sustainable value chains within the local communities for sustainability of the businesses developed (UNWTO 2012).

Conclusion

Developing artist villages as cultural destinations thus supports safeguarding the intangible heritage by augmenting the viability of traditional art forms. It creates income opportunities for both the artists and the larger community by bringing in new patrons. Creative and innovative opportunities for connecting with visitors through cultural exchanges, performances, and tourism services satisfy the aspirations of the communities, especially the youth who have started taking more interest in their local cultural resources. Community-led cultural tourism is thus sustainable when it is conjoined with safeguarding traditional skills, building capacities for heritage management, and ensuring communication channels are established between heritage management departments and public institutions, tourism stakeholders, and communities. Safeguarding of threatened intangible cultural heritage resources through effective management plans, involving local communities and channelling the benefits back to the community, leads to responsible and sustainable tourism development. It is not enough to merely include opportunities of performance or sale and demonstrations of craft-making. Tourism development plans thus need to ensure early and full participation of the local community, including the tradition-bearers, and emphasize promoting and enhancing the viability of local cultural heritage (UNWTO 2012, 86, ICOMOS 1999).

Note

1 Banglanatak dot com (www.banglanatak.com) is the umbrella brand shared by i-land informatics Ltd – a company, and Contact Base – a society.

References

Alma Mater Studioram – Universitàdi Bologna. 2018. *Winter School India 2017* Calls and Program. Accessed June 1, 2018. https://www.unibo.it/sitoweb/c ristiana.natali/contenuti-utili/e81b7ff3

ASEAN. 2016. ASEAN Community-Based Tourism Standard. Jakarta: ASEAN. http://www.asean.org/storage/2012/05/ASEAN-Community-Based-Tourism-Sta ndard.pdf

banglanatak. 2014. *Rural Craft Hub*. Accessed June 1, 2018. http://www. ruralcrafthub.com

banglanatak. 2015. *Art for Life*. Accessed June 1, 2018. http://banglanatak.com/ art-for-life/

banglanatak. 2016. *TourEast*. Accessed June 1, 2018. http://www.toureast.in

banglanatak. 2017. *Rural Craft and Cultural Hubs*. Accessed June 1, 2018. http:// rcchbengal.com

Business Standard. 2019. UNESCO, Rajasthan Govt Join Hands to Promote State's Intangible Cultural Heritage. Accessed May 1, 2020. https://www.business-stan dard.com/article/pti-stories/unesco-rajasthan-govt-join-hands-to-promote-state-s -intangible-cultural-heritage-119090501101_1.html

Clyde, Ellis. 2011. "Reviewed Work(s): Public Indians, Private Cherokees: Tourism and Tradition on TribalGround. Contemporary American Indian Studies by Christina Taylor Beard-Moose." *Journal of Southern History*77(2), 494–495.

International Council on Monuments and Sites. 1999. International Cultural Tourism Charter, Managing Tourism at Places of Heritage Significance. Paris: ICOMOS. http://www.icomos.org/charters/tourism_e.pdf

Levy, Stuart E. and Hawkins, Donald E. 2010. "Peace Through Tourism: Commerce-Based Principles and Practices." *Journal of Business Ethics*89, 569–585.

Onchwati, J., Sommerville, H. and Brockway, N. 2010. "Sustainable Tourism Development in the Masai Mara National Reserve, Kenya, East Africa." *WIT Transactions on Ecology and the Environment*139, 319–330. https://www.wit press.com/Secure/elibrary/papers/ST10/ST10028FU1.pdf

TEDx Talks. June 2015. *Can Music Save the World? | Simon Broughton*. Accessed September 1, 2017. https://www.youtube.com/watch?v=kfQF6fF4kRI (7:30 minutes onwards).

The Indian Express. 2020. "From Bengal, a Patachitra That Narrates the Story of the Fight Against Coronavirus Goes Viral." *The Indian Express*, June 24. https ://indianexpress.com/article/trending/trending-in-india/from-bengal-a-patachitra -that-narrates-the-story-of-the-fight-against-coronavirus-goes-viral-6379099/

Times Now. Dec 2012. Amazing Indians Season 2 Stories Feat Amitava Bhattacharya. Accessed September 1, 2017. https://www.youtube.com/watch?v=jYNMtidTE0c

Tripsavvy. n.d.10Places and *Ways* to *Celebrate* Holi in India. Accessed June 1,2018. https://www.tripsavvy.com/places-to-celebrate-holi-in-india-1539269

UNESCO. n.d.*Intangible Cultural Heritage*. Accessed September 1, 2017. https:// ich.unesco.org/en/lists

UNESCO. 2016. *The Global Report on Culture for Sustainable Urban Development*. France: UNESCO.

UNESCO & banglanatak dot com. 2011. *Indian Heritage Passport Programme on the Cultural Heritage of Purulia: A Concept Paper*. New Delhi: UNESCO. http://unesdoc.unesco.org/images/0019/001928/192865e.pdf

United Nations. 2016. *Sustainable Development Goals*. Accessed September 1, 2017. http://www.un.org/sustainabledevelopment/sustainable-development-goals/

United Nations World Tourism Organization. 2012. *Tourism and Intangible Cultural Heritage*. Madrid: UNWTO.

United Nations World Tourism Organization. 2017. *UNWTO Tourism Highlights 2017 Edition*. Madrid: UNWTO. https://www.e-unwto.org/doi/pdf/10.18111/9789284419029

United Nations World Tourism Organization. 2018. *Tourism and Cultural Synergies*. Madrid: UNWTO.

United Nations World Tourism Organization. 2020. *World Tourism Barometer N°18 January 2020*. World Tourism Organization. Accessed June 24, 2020. https://www.unwto.org/world-tourism-barometer-n18-january-2020.

United Nations World Tourism Organization and the Netherlands Development Organisation. 2010. *Manual on Tourism and Poverty Alleviation – Practical Steps for Destination*. Madrid: UNWTO.

World Bank. 2014. *The World Bank.IBRD.IDA*. Accessed September 1, 2017. http://www.worldbank.org/en/news/press-release/2014/02/03/government-bihar-honors-social-innovators-entrepreneurs-innovation-forum

World Travel & Tourism Council and McKinsey& Company. 2017. *Coping With Success Managing Overcrowded Tourism Destinations*. https://www.wttc.org/-/media/files/reports/policy-research/coping-with-success---managing-overcrowding-in-tourism-destinations-2017.pdf

Yanwei, Han and Zhong, Wei. 2015. *The Role of Cultural Heritage in Economic Generation: The Cases of Shanghai and Hangzhou*. Accessed June 1 2018. IIICH Furnace. Issue 1 (March), https://furnacejournal.files.wordpress.com/2015/06/han-and-zhong-20151.pdf

Past forward

Preparing heritage conservation in India for the 21st century

Ashima Krishna and Manish Chalana

Heritage conservation in India: pressing contemporary concerns

The field of heritage conservation in India has been dominated by the Archaeological Survey of India since the middle of the nineteenth century. Since the 1980s, however, INTACH has spurred on unprecedented private engagement and grassroots efforts at heritage conservation. Countless sites, neighbourhoods, and historic areas have benefitted from this development, and the cadre of professionals and advocates has expanded, along with a growing interest in community participation. Women have always played a critical role in the field, but their involvement has also grown in recent decades. Additionally, allied fields of archaeology, planning, landscape architecture, economic, and community development – among others – are increasingly playing a more important role in heritage conservation, making it truly an interdisciplinary enterprise. There is also increased awareness among communities of what constitutes local heritage, and how they can contribute toward protecting it. Local, state, and central governments have also taken note of these developments, and are attempting to respond with policy changes that allow for more robust conservation and protection frameworks. As the chapters in this volume have shown, both established and emerging professionals are contributing to the evolution and diversification of the field, which bodes well for tangible and intangible heritage in India in the coming decades.

This volume has also presented a range of challenges faced by the field, and the creative ways in which some of those are being addressed in practice. There are many additional continuing challenges as well that need further examination as we assess the future of heritage conservation in India. Here we present a set of ongoing challenges that the field has to grapple with in the 21st century; these are by no means exhaustive, but do provide a glimpse into issues that are worth focusing on for heritage conservation to continue to grow in a more sustainable, equitable, and comprehensive manner. They include: the persistence of an urban and monument-centric heritage framework and its associated problems; unchecked urbanization including government programmes and policies that continue to marginalize heritage conservation; the rise of Hindu nationalism and the glorification of an idealized past along with threats of terrorism on historic sites; growing concerns of climate change and natural hazards with regard to conservation; and critiques of contemporary higher education offerings in heritage conservation. Additionally, we discuss the ASI's continued struggle with managing the country's diverse history and associated heritage. We intend for this discussion to inspire current and emerging professionals as they continue to situate their practice within a rapidly changing urban and professional context.

Beyond the monument

There have been numerous noteworthy developments in India in the last few decades involving governance reform, transparency, and new policy frameworks that have allowed the country to combine its status as the biggest democracy in the world with growing economic prowess. Despite these developments, heritage conservation practice has yet to develop central, state, and local policies that comprehensively engage with and include a range of heritage types beyond the grand monuments and archaeological sites. Even as the mandate of conservation practice has expanded in recent decades, the focus of practice and the bulk of resources are commanded by monuments and sites. Such an approach is both short-sighted and counter-productive (Krishna 2014; Silva and Sinha 2016), and leaves the bulk of unprotected heritage vulnerable to transformation or erasure. As discussed in the introductory chapter, while there have been efforts at preserving non-monumental heritage in the form of historic precincts in some cities like Mumbai, these are currently limited, and need to be expanded across other jurisdictions throughout the country. In recent years, there have been calls for rethinking conservation ideology in the country (Menon 1994, 2003), and creating an Indian approach to cultural landscapes (Thakur 2011, 2007). These calls, however, are yet to be fully realized.

Another avenue of expansion has been the recognition of the context of ASI-designated monuments with the amendment of the Ancient Monuments and Archaeological Sites and Remains Act (AMASR) in 2010. This created prohibited and regulated zones around protected sites of 100 and 200 metres, respectively, with limited results. Much of the immediate context of monuments and sites had already undergone considerable degrees of transformation by the time the amendment went into effect. Several monuments in the country have witnessed their settings transformed over time, and in some instances the transformations themselves could be considered historic in their own right. In managing the historic sites,' contexts, the ASI continues to adhere to the 1923 *Conservation Manual* (Marshall 1923) that recommends creating manicured lawns around historic monuments – an approach that persists today. Sinha calls this approach the "monument-in-the-garden preservation model" (Sinha 2016, 227), which compromises the authenticity of a historic site and presents a contrived image of the setting. There is value in maintaining an authentic context, which lends a greater understanding of a historic site. However, this has not always been successfully achieved in ASI-protected historic sites across the country.

Landscape preservation in India has dwelled mostly on natural landscapes in the form of state and national parks, areas of scenic wonder that were thought of as relatively "pristine" with minimal human contact. In a positive development, heritage conservation has expanded to include historic designed landscapes such as parks and gardens; however, considerable gaps remain in recognizing and protecting sacred and cultural landscapes. Even with designed landscapes that are part of historic sites, especially those from the Mughal era, many have their gardens reimagined as lawns with ornamental planting, completely disregarding the original landscape design or intent. For example, each of the four quads of the quadrilateral garden (*charbagh)* of the Taj Mahal was originally divided into 16 parcels that included hundreds of plants, including fruit and medicinal trees, to create a lush and fragrant environment that gradually revealed the full grandeur of the Taj as one approached it. The flat and formal "lawnscapes" were created in the first decade of the 20th century by Lord Curzon and have since been maintained as such, completely obliterating the original landscape and design intent and context (Herbert 2012). More recently at Warangal, attempts by the ASI to get the Thousand Pillar Temple site included on the World Heritage List involved creating "lawnscapes", which presented a similarly contrived image of the past (Express News Service 2019). The British legacy of "lawnification" that the ASI continues to embrace and promote needs to be problematized as it impacts both the authenticity and integrity of the historic site, not to mention its experience and feel.

Outside of national parks and historic landscapes associated with monuments, the practice of cultural landscape preservation in India is fairly underdeveloped. This is despite the theoretical advances in cultural landscapes in the West as reflected in the works of various preservation agencies and institutions, including the National Park Service and UNESCO. In practice, cultural landscape sites have received only piecemeal attention in India, which is insufficient in retaining the meaning and memory of the site. This is in part due to the complexity of many cultural landscape resources, like the Grand Trunk Road (GTR), which spans across multiple states in India and three countries. The cultural resources associated with this route have been impacted by realignment, demolition, disrepair, replacement, and encroachment (Chalana 2016). While the Punjab segment of the GTR is today on UNESCO's Tentative List, other cultural landscapes in the country have received much less attention if any. Going forward, cultural landscapes would need to be recognized as a resource type and the cultural landscape framework would need to be contextualized for South Asia to capture the full complexity of the multiple and intersecting meanings inherent in Indian cultural landscapes (Thakur 2011).

Integration of heritage conservation with urban development and planning

India has been experiencing a powerful wave of urbanization fuelled in part by globalization. Cities throughout the country are currently participating in a major urban overhaul to upgrade their crumbling infrastructure, create additional housing, and initiate "beautification" projects. With neoliberal reforms infiltrating planning practices, cities are transforming primarily under the dictates of global capital and private sector demands, as well as through the urban programmes furthered by the central government. Urban infrastructure projects, particularly dealing with construction of new metro rail, have had a profound impact on the unprotected historic fabric as well as subsurface archaeology, even in cities like Delhi, which has a greater degree of surveillance (TNN 2015; Verma 2015).

Planning for Indian cities and towns, however, has been criticized for emulating Western ideas in both pre- and post-Independence context (Menon 1997). Contemporary city planning tends to be a relatively mechanical exercise that has resulted in a multitude of development problems in the long term. Even as the masterplans for Indian cities tend to focus more on physical planning and related policy, they fail to effectively integrate the built heritage as part of the planning process. Rapid changes in technology, demographics, and populations also tend to render these documents obsolete within a few years of their inception. In most municipalities, the agencies that create the masterplan, implement that masterplan, approve development projects, implement infrastructure projects, and regulate heritage are all separate, and most often do not coordinate with each other. Consequently, there are a continual series of controversies arising from various kinds of urban development, planning, and heritage conservation projects, with little room for coordination. Heritage conservationists are often involved at a later stage and are projected as obstructionists when they try stalling a planning action that puts heritage at risk.

The implementation of metro rail infrastructure in cities like Pune, Mumbai, Chennai, and Hyderabad has adversely impacted their built and natural heritage primarily due to the bureaucratic dissonance described above (Bari 2018; Bagriya 2018; Hashmi 2019; Tejonmayam 2017). For example, in Chennai, almost 300 buildings, both historic and non-historic, were damaged with cracks during the boring process to create underground tunnels to make way for the metro between 2012 and 2017 (Tejonmayam 2017). While cities like Lucknow have been actively monitoring historic areas where metro construction is underway, lack of mandatory reviews for infrastructure projects to prevent negative impacts to the historic fabric, similar to Section 106 in the United States, has resulted in many losses, including the iconic porch of the historic British colonial-era Divisional Railway Manager (DRM) Office, located in the heart of old Lucknow, in 2018 (TNN 2018a; Srivastava 2019).

Urban renewal programmes are primarily tasked with addressing challenges associated with rapid urbanisation including poverty alleviation; most do not expressly deal with historic areas of urban India. Even the ones that have, have underperformed and not fully addressed or resolved the challenges posed by heritage conservation in the context of rapid urban development. For example, the Jawaharlal Nehru National Urban Renewal Mission (JNNURM), a ten-year nationwide project launched in 2005, identified 63 pilot cities from across the country to implement integrated management and development, including revitalizing the historic urban cores of Indian cities. While hailed as one of the most ambitious projects of its time (Kundu 2014, 618), JNNURM's record on furthering heritage conservation was limited, demonstrating the inability of the central government to properly motivate local governments to create capacity in order to accomplish approved projects (Naik 2012).

In 2015, with the advent of new leadership at the central level, the 500-crore (crore = ten million) Heritage City Development and Augmentation Yojana (HRIDAY) was launched with 12 pilot cities (completion date November, 2018) from across the country including Ajmer and Amritsar (Capital Market 2018). Several years into the project, it is becoming clear that there is a need for better inter-agency cooperation and management for the projects to be successful. In Amritsar, for example, several sites (including the Ram Bagh Gate discussed in this volume) that were successfully conserved by the municipal corporation under the auspices of HRIDAY were taken out of the public domain for extended periods of time (Teja 2019a). While the Ram Bagh museum was eventually opened to the public in May 2019 (Teja 2019b), the institutional chaos in the aftermath of conservation highlights the often dysfunctional and dissonant nature of various government agencies. State and local administration in Puri, on the other hand, failed to meet project deadlines for six projects undertaken as part of HRIDAY, and eventually lost access to additional funding, which ended in November 2018 (Mishra 2018). To put HRIDAY in perspective, it accounted for less than one percent of the total federal allocations for various kinds of development projects across the country (Jadhav 2019). Even with its successes in some areas, the programme has not greatly transformed the practice of heritage conservation in the country.

Urban heritage is also being impacted by state-sponsored vandalism and erasure masquerading as "urban revitalization", particularly when seen through the lens of emerging Hindu nationalism in the country. Varanasi, the epicentre of Hindu religion and spirituality, for example, has recently seen rapid loss of historic fabric. In 2018, almost 300 houses were demolished

Figure 18.1 Politically motivated demolition and transformation in the heart of Varanasi has impacted several vernacular and historic neighbourhoods, not just eroding their intrinsic character and streetscape but also displacing the residents. Source: "Varanasi – INDIA" by polosopuestosblog is licensed under CC BY 2.0

in some of the oldest parts of the city (Figure 18.1) to make way for tourist and pilgrim facilities and to create a sanitized space around the iconic Kashi Vishwanath Temple (Ghosh 2018; Agarwal 2019; Srivathsan 2019). The project has a distinct political emphasis, particularly in its connections to the national political party in power in 2018 (Srivathsan 2019). This form of spatial cleansing also points to the vulnerability of the unprotected historic fabric of a city that has both historic value and social meaning for the residents. The Ministry of Tourism's efforts at augmenting the tourism and visitor potential of religious cities like Varanasi also led to the launch of the programme Pilgrimage Rejuvenation and Spirituality Augmentation Drive (PRASAD) in 2015. While this programme would have been commendable given India's religious plurality, unsurprisingly, the majority of the sites selected for the programme were significant to the Hindu faith. Additionally, over the years, Varanasi has witnessed many efforts by academics, scholars, INTACH, and local, state, and central government agencies to effectively manage its heritage. There have also been local efforts at designating the area as a World Heritage Site (R. P. B. Singh and Rana 2017; R. P. B. Singh 2016). The waterfront and *ghats* have also been part of several plans, including masterplans in 1991 and 2011, a City Development

Plan in 2006 for JNNURM, the recent Smart City[1] plan, the aforementioned HRIDAY scheme, and the City Development Plan for 2041, among others (Srivathsan 2019). As with many other cities, however, lack of local legislation and policy, as well as absence of political will has rendered these plans mere documents. Publications from local scholars also suggest conflict among academics, activists, and INTACH over projects' vision, goals and implementation (R. P. B. Singh 2016; R. P. B. Singh and Rana 2017).

Vernacular, non-urban, and intangible heritage

India is a repository of everyday built fabric associated with various periods of historical development that continues to serve the populations that live and work in these neighbourhoods. However, vernacular sites remain unrecognized and ignored by local, state, and central agencies. Efforts to conserve non-monumental vernacular and/or living heritage have typically come from some larger municipalities such as Mumbai and local advocacy groups concerned about changing neighbourhoods. INTACH has recently created a *Charter for the Conservation of Unprotected Architectural Heritage and Sites* in India to shape a practice that is more inclusive of vernacular and recent heritage. However, most communities either undervalue or lack resources and are inefficient in protecting their vernacular heritage.

There has been some success in cities like Ahmedabad and Cochin. Ahmedabad took stock of its vernacular heritage comprising *pols* and *havelis*, and created a municipal programme in the 1990s to conserve and revitalize the walled city and its vernacular fabric (Nayak 2016, 215). The local Ahmedabad Municipal Corporation (AMC), NGOs, and activists have since worked on various initiatives through public–private partnerships. For example, the Government of France, in collaboration with the AMC and HUDCO, helped execute the restoration of vernacular heritage in the city's historic core through the local government's Heritage Cell (Nayak 2016, 215). In 2001, the city also created first-of-its-kind clauses in town planning legislation, which prohibited any historic property from being demolished without a permit from the city's Heritage Cell. In 2007 they introduced heritage regulations for the city to monitor and enforce restrictions on any changes to historic properties in the city, which have proved to be a model set of regulations for other municipalities in the state (Nayak 2016, 216). All these efforts paid off in 2017 when the walled city of Ahmedabad was inscribed as world heritage by UNESCO – a first such designation for an Indian city.

Similarly, non-urban heritage continues to receive little attention, even in emerging suburban and exurban locales like the Gurgaon District in Haryana (P. G. Munjal 2016; P. Munjal and Munjal 2017), which is part of the larger National Capital Region of New Delhi. While urban areas may struggle with administrative apathy and limited funding, the issues are further amplified in small towns and villages where basic amenities like electricity and water are often scarce – heritage and its conservation therefore is often a non-priority. With the exception of monuments under state ASI jurisdiction and efforts by INTACH documenting cultural resources in small towns, much of this heritage remains neglected and unprotected. Going forward, heritage conservation will need to expand beyond the urban to non-urban settings to create local policies to protect heritage and ensure compliance with national- and state-level policies for ASI-listed monuments and sites.

For heritage conservation practice to be more effective, it will need to integrate better with intangible heritage in the decades to come. Often intangible heritage is located in non-urban settings, so reviving the traditional cultural practices, art forms, and tourism can offer livelihood opportunities to residents. Understanding, documenting, and promoting intangible heritage can bring about significant social and environmental changes by promoting at-risk livelihoods, practices, and traditions and bring them into the mainstream (Sinha 2016, 229). Conservation professionals will therefore need to continue to pay attention toward recognition and integration of intangible and tangible heritages. Care should be taken to not dislocate the intangible heritage from its original setting as a means of reviving it. This would ensure that the crafts' context and populations that have kept it alive continue to benefit from it. Several efforts are currently underway to ensure the continuity of intangible heritage in its original context is maintained in West Bengal and Punjab as discussed in this volume. Similar efforts should be expanded into different contexts and integrated with livelihoods to maintain continuity and relevance of intangible heritage.

The threat of rising Hindu nationalism

In recent decades, India has seen rising Hindu nationalism, long before the controversial Varanasi redevelopment. The most iconic example of claiming Hindu legacy for Muslim sites is the Babri Masjid in Ayodhya. Since the 1940s, the 16th-century mosque has been a site of dispute: Hindu devotees have claimed that the mosque was built on the site of a temple that marked the birthplace of Lord Rama (for more on the history of the site see Paola and Bacchetta 2000; Islam 2007). Excavations by archaeologists in the 1860s and again in the 1960s suggest Buddhist connections to the site as well (Islam 2007). In a bid to drum up nationalistic fervour, however, a politically motivated rally (*rath yatra*) in 1992 led to the senseless demolition of the site, and further led to communal riots across the country.

Archaeologically and historically, the site has also been contested, and has been under litigation for decades.[2] The ASI was tasked with undertaking excavations to help the courts decide on the case; however, the agency's fairness has been repeatedly called into question for their 2003 report alluding to the presence of alleged temple pillars and structures "1.5 feet (0.46 metre) below the lowest floor of the mosque..." (Islam 2007; Abbas 2004;

Sharma 2018). Independent archaeologists have, in fact, called out the ASI for repeatedly misrepresenting information to suit the nationalist narrative of the presence of a temple below the mosque (Sharma 2018).

In 2020, the current central government has circled back to their playbook from the 1990s, and nationalism and Hindutva are again making their presence felt in all walks of life. Without a doubt, the Babri Masjid site, as well as other historic sites belonging to minority groups, may again come under other such threats. Similarly, many cities and sites are now being renamed to reclaim Hindu identity. While initially the renaming of cities was done to shed a painful colonial past, more recent renaming efforts, primarily in conservative-leaning states, point to an effort at erasing the Islamic past and imposing an idealized Hindu identity on place (TNN 2018b). In 2017, for example, the pro-Hindutva chief minister of Uttar Pradesh drew significant criticism when he alluded to the Taj Mahal as not being India's heritage because of its association with Islam and Muslims (Qureshi 2017). This kind of state-sanctioned de-legitimization of heritage can have devastating impacts on other, lesser known sites that do not enjoy the same prestige and world-wide recognition as the Taj Mahal. The heritage conservation profession would need to remain true to its democratic roots and guard India's constitutional pluralism to recognize and affirm its cultural diversity, and ensure that practice does not privilege one group over the other. This would help the field expand its constituent base and reaffirm minority groups' faith in the profession.

Global conflict and terrorism

Even as Babri Masjid's destruction can be seen as an act of domestic terrorism, threats from war and global terrorism are becoming increasingly pressing all over the world, and the Indian subcontinent is no exception. Historic and iconic sites like the Taj Mahal Palace Hotel, which celebrated 115 years in December 2018, are today not just a symbol of eclectic architectural influences (AD Staff 2018), but are essential records of urban historical development. The Taj Mahal Palace Hotel was built at the turn of the 20th century, and opened for business in 1903. In November 2008, it was one of several sites across Mumbai that came under terrorist attack. Over the four-day siege, the iconic building suffered extensive damage, in addition to tragic loss of lives. While some parts of the hotel reopened to the public within a few weeks of the attack, the palace wing and other sections underwent a two-year restoration (Deepak 2018).

Since the devastating attack, the hotel has been fully restored, but threats to similar iconic heritage sites across the country still remain. As history has shown, in times of conflict, the most effective targets are often heritage sites or places of worship (as in the case of the Bamiyan Buddhas demolished by the Taliban in 2001). The threats of global terrorism are therefore a credible and urgent concern that heritage conservation professionals would need to address in their mitigation strategies for historic sites. In extreme cases where demolition and destruction does occur as an act of terror or war, professionals will also need to recognize and appreciate the altered spatial and historical meaning of the damaged site, and design a conservation approach based on the new context. This can range from trying to interpret the void caused by the destruction of the majestic Bamiyan Buddhas in Afghanistan, to the appreciation and recognition of the cultural significance of the wholly reconstructed and replicated Stari Most bridge in Mostar in Bosnia and Herzegovina. International organizations like UNESCO and the World Heritage Centre are increasingly expanding their mandate to encompass these altered meanings; thus, despite destruction and reconstruction respectively, both the Bamiyan Buddhas and the Stari Most bridge sites are recognized as world heritage. These examples might offer lessons for the Babri Masjid site's reconstruction to interpret its violent history and offer contemporary opportunities for healing.

The ASI: ongoing challenges

At the central level, ASI's role as the custodian of India's iconic heritage has been undisputed for over 150 years. Today, however, the ASI also continues to operate within relatively unchanged frameworks including outdated lists of national monuments, archaeological sites, and artefacts. For conservation practitioners, there is a dearth of resources available to update ASI's often underdeveloped surveys of historic sites. Consequently, there has been minimal effort at expanding the list; thousands of architecturally, historically, and archaeologically significant sites are thus languishing and fast disappearing due to lack of protection, designation, and enforcement at federal and state levels (Chainani 2007a, 2007b; Rao 2018). Going forward, comprehensive listing and documentation would need to be a priority; otherwise loss of unlisted historic sites would continue unchecked.

Additionally, over the years, the agency has been mired in several controversial decisions that have impacted its reputation. Most recently, in 2012, the ASI was involved in examining the remains of a 17th century mosque, the Akbarabadi Masjid, found at Subhas Park during Delhi Metro's third phase of construction (Ali and Ashok 2012). A local resident tried to construct a wall and minarets at the site where the building remains were found – he then filed a suit in the Delhi High Court when the ASI objected to the construction. The ASI in turn dismissed the significance of the building remains found, and consequently, the courts allowed the Delhi Metro Rail Corporation (DMRC) to commence construction after a two-year halt (Indo-Asian News Service 2013; Anand 2014). This has had a long-lasting and irreversible impact on the archaeological and architectural resources in the area, and sheds a negative light on the agency.

Ultimately, the ASI struggles with limited governmental appropriations to effectively carry out its work, which impacts its ability to expand the investigation of unexplored sites and invest in the proper maintenance of listed sites. Given the preponderance of historic sites in the country, many are not protected and routinely compromised by incongruous additions and alterations, or worse, demolition. Lack of financial resources has also led to a relatively new phenomenon of "privatization" of the maintenance and upkeep of national heritage. In 2018, for example, the Dalmia Bharat Group signed a memorandum of understanding with the Ministry of Tourism and the ASI to "adopt" the iconic Red Fort in Delhi for five years under the government's "Adopt a Heritage Scheme". The government is actively seeking calls from other public and private entities to similarly "adopt" sites like the Taj Mahal, Kanheri Caves, Sun Temple in Konark, and Agra Fort (Shrivastava 2018). This move has garnered criticism and concern, prompting fears that other major historic sites may be similarly taken over by private organizations under the guise of adoption for care and maintenance, resulting in increasing private uses, and compromised public access (Patwardhan 2018).

In recent years, the ASI has also struggled with maintaining the quality of repairs and restoration carried out both by other stakeholders at national monuments, and by the agency itself. Most recently, citizen activists brought to public attention the abysmal quality of restoration work at Mumbai's iconic Indo–Saracenic style Victoria Terminus (now called Chhatrapati Shivaji Maharaj Terminus, or CSMT; and a World Heritage Site). Central Railways, the principal custodian of the structure, did not hire trained conservation architects to carry out the restoration work, which resulted in a caricature of the iconic sculptures on the building's façade (TNN 2019). In Lucknow, the restoration work carried out by ASI staff at the iconic Kaiserbagh gates has significantly altered the delicate stucco reliefs, compromising their authenticity and appearance (Ashima Krishna, site visit, 2 March 2012). The ASI has also often made a mockery of its own codes and legislation. In March 2019, for example, the ASI was criticized for allowing and undertaking construction of buildings and a golf course in and around the iconic Golconda Fort in Hyderabad. Local activists have heavily criticized the scale of activity being undertaken by the ASI, and the irreversible damage it could bring to the historic site (Turaga 2019). In the coming decades, the ASI would need to reassess its values and approaches to conservation, to engage contemporary approaches to practice, and align better with the international discourses and best practices.

Heritage conservation, changing climate, and natural hazards

Concerns of climate change are also increasingly pressing in the Indian subcontinent. Changes in climatic patterns, particularly typhoons, floods, fires and possibly earthquakes pose a risk to heritage sites across the country, but particularly in hazard-prone regions. Devastating disaster events in the last few years in Gujarat, Uttarakhand, Odisha, Tamil Nadu, and Maharashtra, among others, have shown just how vulnerable our towns, cities, and villages are to loss of life and property. Conservation professionals are now increasingly aware of the impacts of natural disasters and climate change on the built environment (Jigyasu 2019); however, many historic sites have yet to be retrofitted for seismic hazards or have an emergency evacuation plan in place. Even as traditional buildings in different hazard-prone zones have shown to perform better in the event of a natural disaster (compared to contemporary buildings) they are routinely replaced by poorly built modern construction without seismic provisions. Hazard mitigation plans for these areas should recognize the value of traditional building systems and integrate heritage conservation as part of their plan. Major monuments in the country are addressing this challenge; however, going forward, hazard mitigation should inform all heritage conservation practice in hazard-prone areas across a range of sites including cultural landscapes and intangible heritage.

In addition to heritage sites impacted by natural disaster, heritage conservation practice in India would have to prepare for climate change, particularly rising sea levels and receding glaciers. While environmental scientists have been aware of the problems of climate change for a considerable time, conservationists have only recently begun to address the issue in connection with cultural heritage. In the Himalayan region, climate change, marked locally by shifting monsoons, increasing landslides and retreating glaciers, is a serious threat to the local population, tourism, and traditional heritage. In coastal regions, typhoons, cyclones, and floods pose a direct risk to life and property. There are currently few professionals and academics that are exploring the relationship between conservation and environmental concerns in the context of India. Going forward, there will need to be a more systematic focus on mitigating the impacts of environmental events, natural hazards, and climate change actions on cultural heritage sites and landscapes.

Another continuing challenge to the heritage conservation field deals with advocacy and activism for both environmental and heritage issues. In the absence of policies and/or enforcement, activism has played a huge role in India. At federal, state, and local levels, lack of funding is most often the reason cited for lack of enforcement of policies already in place. In many cases, advocates and activists including the late Shyam Chainani in Mumbai, and Debashish Nayak in Ahmedabad, have had significant impact in pushing local governments to create policies and legislation that designate, list, protect, and preserve historic sites and areas (for more on this topic see Nayak 2016; Nayak and Iyer 2008; Chainani 2007a, 2007b). In other cases, advocates like Mohammad Haider in Lucknow have sought judicial assistance and oversight in the absence of a municipal response. More recently, in December 2018, another citizen advocate in Ahmedabad, Munaf Ahmed Mullaji, filed a public interest litigation (PIL) in the Gujarat High Court to have the court direct the ASI to take criminal action against the divisional railway manager (DRM) who undertook illegal construction within the prohibited 100-metre zone of a nationally protected monument, Brick Minarets (TNN 2018c). This came about after the ASI twice sent notices to the Department of Railways, once in April and again in July of 2018. Lack of inter-agency cooperation, however, eventually led to the involvement of the judiciary via citizen advocates. Public interest litigation is now a popular tool for local advocates to fill in the gap often created by lack of enforcement and criminal prosecution by the ASI against those who flout federal rules and regulations. But it also points to a major inability of an agency like the ASI to carry out its most important mandate: that of safeguarding nationally protected "monuments". The profession and conservation education therefore have an opportunity to create spaces to nurture a range of activism that would lead to public support and conservation. A good contemporary example is the Sushant School of Art and Architecture's studio for the Master of Architecture in Built Heritage programme. As part of the educational process, students undertake a 100-day challenge to highlight issues related to "lost typologies" of the built environment. The latest iteration of the exercise focused on Kaman Sarai, a significant building in Gurgaon, through workshops, documentation, condition assessment, and developing proposals for the site (Ansal University 2019). This interactive exercise allows students to engage with activists and community and preservation stakeholders to share and learn critical skills of practice.

Equity and social justice

Another major challenge for heritage conservation going forward revolves around equity and justice. As mentioned earlier, current federal, state, and local policies often tend to sanitize historic sites and areas, with detrimental effects on the poor and marginalized living or making their livelihood in and around a historic site or area. One of the most iconic world heritage sites in the country, Hampi, has struggled with such issues. In 2011, for example, the Bellary district administration obtained an order from the Karnataka High Court to evict the residences and demolish decades-old shops and homes fronting the iconic (and functioning) Virupaksha Temple. The contemporary bazaar was run by local entrepreneurs who catered to visitors and pilgrims, and was a thriving hub of activity. For local administrators, however, the illegal encroachments, though decades old, were an eyesore, and thus the clandestine eviction and demolition activity caught world-wide attention because of the aggressive way in which it was carried out (Ravindran 2011; Equitable Tourism and March 2011). More recent developments in Varanasi show similar attempts at cleansing and sanitizing the "living" and "messy" aspects of historic areas that impact low-income populations disproportionately.

Elsewhere in the country, while the development of a historic precinct framework to protect neighbourhoods is a step in the right direction, it does not engage diverse neighbourhoods of different socioeconomic statuses. Currently all historic precincts in Mumbai, for example, engage the history and memory of the elite in the city. The working-class neighbourhood of Girangaon, considered the birthplace of the textile industry in India (and nick-named "Manchester of the East") was not listed as a historic precinct, which has ultimately led to its transformation into a high-end neighbourhood show-casing hypermodern and global aesthetics (Figure 18.2) without much regard to its working-class history and heritage (Chalana 2009). The expansion of current conservation practice in India continues to emulate models that are insufficient in appreciating low-income neighbourhoods where lifeways, live-lihoods, and built environments are inextricably linked, and remain sites of "conflicting claims about the role of history, and of particular visions of his-tory" within the framework of globalizing Indian cities (Chalana 2009, 2).

Gentrification has also impacted other heritage precincts includ-ing Panaji (Goa), New Delhi, and Cochin. In each, working-class

Figure 18.2 A view of the Phoenix Mills in Lower Parel, Mumbai, which is now a shopping mall. Over the years, most vacant mill sites have been sold off for redevelopment, or worse, demolition, thus erasing a very important part of Mumbai's historical development as an economic powerhouse. The role played by the mill workers in that process is also often ignored. Source: Rakesh Krishna Kumar; originally posted to Flickr as Phoenix Mills, CC BY-SA 2.0, https://commons.wikimedia.org/w/index.ph p?curid=5876464

neighbourhoods have been co-opted into elitist zones catering to affluent tourists in the form of boutiques and galleries, heritage hotels and restaurants. Although this may prevent excessive destruction of historic fabric, it has occurred without much concern for the displacement of residents and still yields several negative impacts on historic fabric as well. While many older buildings have been restored and adaptively reused, others have been altered beyond recognition; many others have been demolished and replaced by contemporary structures with much of the new uses catering to the tourist economy. As elsewhere in India, gentrification is accepted as an unintended outcome of conservation practice, but that fatalistic view needs to be challenged with creative approaches where revival of a neighbourhood does not always have to rely on tourist economy and lead to changing land uses and demographic shifts. Social justice and equitable urban development need to become central issues for conservation professionals going forward.

Education

Preservation education is another factor that needs serious consideration. As of 2020, India had 32 graduate programmes in archaeology, seven graduate programmes in architectural conservation,[3] seven graduate programmes offering dual concentrations in conservation and an allied field like museology, three graduate programmes in heritage management, and three programmes in heritage and tourism management at various universities and institutions. With only seven programmes in architectural conservation, there is still a paucity of topical higher education in the country, and the current system limits the ability of students to take courses across disciplines. As a consequence, students in architecture, planning, and archaeology mostly take coursework in their own fields, and therefore may not always be well-versed in tackling contemporary conservation challenges through interdisciplinary lenses. Many foreign-trained professionals have a relatively broader exposure to different kinds of courses, topics, faculty, and projects, which better prepares the professionals to deal with the complexity of conservation projects.

Another issue with conservation education in India is that most contemporary conservation programmes are housed in departments of architecture, making their focus distinctly design- and built-environment-based. In the past few decades, there have been calls for reassessing both architectural and conservation education (Menon 1999; Thakur 1990; Mehrotra 2007), since the majority of conservation professionals in India are trained as architects. More recently, Mehrotra has critiqued conservation education in India as not preparing "potential practitioners to intervene in the Indian urban context" (Mehrotra 2007, 349). The emphasis on materials and design in conservation training programmes, according to Mehrotra, ignores the dynamism found in the everyday performances taking place within what he has termed the "kinetic city" (Mehrotra 2007, 342). The concept of the kinetic city might offer lessons to expand the notion of intangible heritage to integrate it more effectively with the "static city".

There have been some recent developments that have tried to address these critiques. The INTACH Heritage Academy (IHA), for example, offers training courses and programmes for agencies, departments, artisans, and professionals. The IHA also has a year-long post graduate diploma course, which is connected to the graduate programme at York University in England, and offers ten diverse modules that show a cross-disciplinary approach to heritage conservation. Even though these courses do not result in a graduate degree, they are useful for students and practitioners who want to develop expertise in specific conservation topics. In the coming decades, however, as heritage conservation in India becomes even more specialized, students and practitioners would need to be trained in both traditional and non-traditional ways to meet the growing demands of the profession. Such valid critiques of conservation education (including those not discussed here) need to be addressed to ensure that conservation professionals are fully equipped to meet the changing contexts, needs, aspirations, values, and expectations in the 21st century.

Looking ahead

A conservation professional or an activist should not only be concerned with how to treat the remnants of the past, but also be "one who gives expression to contemporary aspirations" (Mehrotra 2016, xviii). Looking ahead, it will be imperative for conservation practitioners to understand how contemporary urban development is shaped, and how in turn it shapes the historic core of urban India. Today there is a range of restoration and conservation practices found across the country, but there is no central group or organization that mandates and regulates the conservation field through education and professional practice. Consequently, there is currently no accountability in either the public or private sector. In the future, systems would have to be created to ensure the standardization of practice, to help guide emerging professionals and to discourage arbitrary, and often egregious conservation practices carried out by those not familiar with the field. Consistent sets of rules, guidelines, and standards for conservation in India would go a long way towards rationalizing and streamlining conservation practice. The technical manuals being developed by INTACH for different cultural resources and material types are a step in the right direction, but this effort needs to expand considerably to include the full range of sites and materials.

This volume has discussed the many challenges facing the public sector, while also identifying new best practices in conservation that have often emerged from the private sector. Conservation professionals have paved unique pathways in heritage conservation practice, and more emerging professionals will continue to diversify the field. Ultimately, for heritage conservation to become effective in protecting a full range of India's diverse heritages, it would need to have legislative backing at all levels of government and reasonable funding to carry out the work, along with the rules, guidelines, and standards to promote best practices.

Notes

1 Government of India's Smart City initiative has aimed to bring comprehensive planning and development to Indian cities, with mixed results. Most notably, while the programme acknowledges the inherent ambiguities in trying to define a "smart city", it fails to account for any type of heritage (built, intangible, archaeological) in its approach, while focusing primarily on greenfield development. For more on the Smart City initiative, read here: http://smartcities.gov.in/content/innerpage/smart-city-features.php
2 In November 2019, the Supreme Court of India finally issued a verdict after decades of litigation—in favour of the Hindu litigants—allowing them to construct a Hindu temple at the disputed site. The Court also ordered the central government to allot a five acres of the site to the Sunni Waqf Board for the construction of a mosque (Mathur 2019).
3 The first graduate program in conservation was initiated by the School of Planning and Architecture, New Delhi in 1987, and it remained the only such program for almost three decades until the Sinhgad College of Architecture launched a program in 2006.

References

Abbas, Rizvi Syed Haider. 2004. "Taking ASI by the Horns." *The Milli Gazette*, April. http://www.milligazette.com/Archives/2004/16-30Apr04-Print-Edition/1604200406.htm

AD Staff. 2018. "EXCLUSIVE: The Taj Mahal Palace Like You Have Never Seen It Before." *Architectural Digest*, July 21. https://www.architecturaldigest.in/content/taj-mahal-palace-hotel-mumbai-photos/#s-cust0

Agarwal, Kabir. 2019. "In Modi's Varanasi, the Vishwanath Corridor Is Trampling Kashi's Soul." *Wire*, March 8. https://thewire.in/politics/kashi-vishwanath-corridor-up-bjp

Ali, Mohammad, and Sowmiya Ashok. 2012. "Remains of 17th Century Mosque Discovered Near Delhi Metro Corridor." *The Hindu*, July 7. https://www.thehindu.com/news/cities/Delhi/remains-of-17th-century-mosque-discovered-near-delhi-metro-corridor/article3612859.ece

Anand, Utkarsh. 2014. "SC Lets Metro Begin Work Near Disputed Delhi Mosque." *Indian Express*, September 9. https://indianexpress.com/article/cities/delhi/sc-lets-metro-begin-work-near-disputed-delhi-mosque/

Ansal University. 2019. "Save the Sarai –100-Day Challenge." Snapshot 2019 2nd Quarter. https://ansaluniversity.edu.in/ansal-snapshot-2nd-quarter/

Bagriya, Ashok. 2018. "SC Allows MMRCL to Continue Metro-3 Work below Mumbai Fire Temples." *Hindustan Times*, December 10. https://www.hindustantimes.com/mumbai-news/sc-allows-mmrcl-to-continue-metro-3-work-below-mumbai-fire-temples/story-7VMGhnYbKWNHE0bmean1nL.html

Bari, Prachi. 2018. "Pune Metro States No Option, to Go through Salim Ali Bird Sanctuary." *Hindustan Times*, December 26. https://www.hindustantimes.com/pune-news/pune-metro-states-no-option-to-go-through-salim-ali-bird-sanctuary/story-jpsCap509LhJJpoeA48qfI.html

Capital Market. 2018. "More Than Rs.240 Crore Released for 12 Cities under HRIDAY Scheme." *Business Standard*, March 6. https://www.business-standard.com/article/news-cm/more-than-rs-240-crore-released-for-12-cities-under-hriday-scheme-118030600390_1.html

Chainani, Shyam. 2007a. *Heritage & Environment: An Indian Diary*. Mumbai: Urban Design Research Institute.

Chainani, Shyam. 2007b. *Heritage Conservation, Legislative and Organisational Policies for India*. New Delhi: INTACH.

Chalana, Manish. 2009. "Of Mills and Malls: The Future of Urban Industrial Heritage in Neoliberal Mumbai." *Future Anterior* 9(1): a-15. https://muse.jhu.edu/journals/future_anterior/v009/9.1.chalana.html

Chalana, Manish. 2016. "'All the World Going and Coming': The Past and Future of the Grand Trunk Road in Pubjab, India." In *Cultural Landscapes of South Asia: Studies in Heritage Conservation and Management*, edited by Kapila D Silva and Amita Sinha. Routledge. 92–110.

Deepak, Vineeta. 2018. "10 Years on, Mumbai Moves on from Attacks but Scars Remain." *AP News*, November 23. https://www.apnews.com/22586837d4054fa4820d37ea67117a0d

Equitable Tourism, and Equations March. 2011. "Inhuman and Illegal Eviction and Demolition at Hampi Bazaar." July: 1–11. http://www.equitabletourism.org/stage/files/fileDocuments1076_uid18.pdf

Express News Service. 2019. "Efforts on to Get Heritage Tag for 1000-Pillar Temple in Warangal." *The New Indian Express*, February 26. http://www.newindianexpress.com/states/andhra-pradesh/2019/feb/26/efforts-on-to-get-heritage-tag-for-1000-pillar-temple-in-warangal-1943793.html

Ghosh, Bishwanath. 2018. "Beautification Plan Destroys Oldest Neighbourhoods in Varanasi." *The Hindu*, December 9. https://www.thehindu.com/news/national/other-states/beautification-plan-destroys-oldest-neighbourhoods-in-varanasi/article25704389.ece?utm_campaign=article_share&utm_medium=referral&utm_source=whatsapp.com

Hashmi, Rasia. 2019. "Metro in Old City: Notices Issued to Acquire Properties." *The Siasat Daily*, February 16. https://www.siasat.com/news/metro-old-city-notices-issued-acquire-properties-1467978/

Herbert, Eugenia. 2012. "Curzon Nostalgia: Landscaping Historical Monuments in {India}." *Studies in the History of Gardens & Designed Landscapes* 32(4): 277–296. https://doi.org/10.1080/14601176.2012.719715.

Indo-Asian News Service. 2013. "No Akbarabadi Mosque Ever Existed, Apex Court Told." *Zee News*, February 26. https://zeenews.india.com/news/delhi/no-akbarabadi-mosque-ever-existed-apex-court-told_831777.html

Islam, Arshad. 2007. "Babri Mosque: A Historic Bone of Contention." *The Muslim World* 97(2): 259–286. https://doi.org/10.1111/j.1478-1913.2007.00173.x

Jadhav, Radheshyam. 2019. "Urban Development: Majority of Centre's Assistance to States Goes to Metro Rail, Housing Projects." *The Hindu Business Line*, January 4. https://www.thehindubusinessline.com/news/national/urban-development-majority-of-centres-assistance-to-states-goes-to-metro-rail-housing-projects/article25901861.ece

Jigyasu, Rohit. 2019. "Managing Cultural Heritage in the Face of Climate Change." *Journal of International Affairs* 73(1): 87–100.

Krishna, Ashima. 2014. *The Urban Heritage Management Paradigm: Challenges from Lucknow, An Emerging Indian City*. Cornell University. http://ecommons .library.cornell.edu/handle/1813/36044

Kundu, Debolina. 2014. "Urban Development Programmes in India: A Critique of JnNURM." *Social Change* 44(4): 615–632. https://doi. org/10.1177/0049085714548546.

Marshall, John Hubert. 1923. *Conservation Manual. A Handbook for the Use of Archaeological Officers and Others Entrusted with the Care of Ancient Monuments*. Calcutta: Superintendent Government Printing India.

Mathur, Aneesha. 2019. "Supreme Court Dismisses All Ayodhya Verdict Review Pleas." *India Today*, December 12. https://www.indiatoday.in/india/story/ay odhya-verdict-review-petitions-ram-mandir-supreme-court-1627737-2019-12-12

Mehrotra, Rahul. 2007. "Conservation and Change: Questions for Conservation Education in Urban India." *Built Environment* 33(3): 342–356. https://doi. org/10.2148/benv.33.3.342

Mehrotra, Rahul. 2016. "Prologue: Imagining the Future of Conservation in South Asia." In *Cultural Landscapes of South Asia: Studies in Heritage Conservation and Management*, edited by Kapila D. Silva and Amita Sinha. Abingdon: Routledge, xvii–xix.

Menon, A.G. Krishna. 1994. "Rethinking the Venice Charter: The Indian Experience." *South Asian Studies* 10(1): 37–44. http://www.tandfonline.com/doi /abs/10.1080/02666030.1994.9628475.

Menon, A.G. Krishna. 1997. "Imaging the Indian City." *Economic and Political Weekly* 32(46): 2932–2936. https://doi.org/10.2307/4406063.

Menon, A.G. Krishna. 1999. "Reforming Architectural Education." *Architecture + Design*.

Menon, A.G. Krishna. 2003. "The Case for an Indian Charter." In *Seminar*, edited by Tejbir Singh, 48–53. New Delhi: Malvika Singh. http://scholar. google.com/scholar?hl=en&btnG=Search&q=intitle:The+case+for+an+Indian+ Charter#1

Mishra, Sandeep. 2018. "Puri Lacks Heart in Hriday Scheme." *The Telegraph*, July 21. https://www.telegraphindia.com/states/odisha/puri-lacks-heart-in-hriday-sch eme/cid/1419388.

Munjal, Parul G. 2016. "Construction of Heritage: Small and Medium Towns of Gurgaon District." *Journal of Heritage Management* 1(2): 98–125. https://doi. org/10.1177/2455929616682079.

Munjal, Parul G., and Sandeep Munjal. 2017. "Built Heritage in Small Towns a Unique Tourism Opportunity: Case of Shiv Kund, Sohna." *Journal of Services Research* 17(2): 17–40.

Naik, Mukta. 2012. "Review Lessons from JNNURM." *Context* 9(1): 93–96.

Nayak, Debashish. 2016. "Getting the City Back to Its People: Conservation and Management of Historic Ahmedabad, India." In *Cultural Landscapes of South Asia: Studies in Heritage Conservation and Management*, edited by Kapila D Silva and Amita Sinha. Abingdon: Routledge, 211–226.

Nayak, Debashish, and Anand Iyer. 2008. "The Case of Ahmedabad: Heritage Regulations and Participatory Conservation." *Context: Built, Living and Natural* 5(1): 175–182.

Paola, Bacchetta. 2000. "Sacred Space in Conflict in India: The Babri Masjid Affair." *Growth and Change* 31(2): 255–284. https://doi.org/10.1111/0017-4815.00128.

Patwardhan, Sujit. 2018. "Looking beyond Red Fort : Is Privatisation the Only Way to Save Our Monuments ?" *Citizen Matters*, May 8. http://citizenmatters.in/adopt-a-heritage-government-scheme-red-fort-dalmia-6472

Qureshi, Siraj. 2017. "Yogi Adityanath's Statement on Taj Mahal Irks People of Agra." *India Today*, June 30. https://www.indiatoday.in/india/story/yogi-adityan ath-taj-mahal-agra-mughals-1021709-2017-06-30

Rao, Mohit M. 2018. "Awash with History, yet Neglected." *The Hindu*, December 8. https://www.thehindu.com/news/cities/bangalore/awash-with-hi story-yet-neglected/article25696481.ece

Ravindran, Nirmala. 2011. "Consigned to the Debris of History." *India Today*, August.

Sharma, Betwa. 2018. "There Is No Evidence of A Temple Under the Babri Masjid, Just Older Mosques, Says Archeologist." *Huffington Post* India, May 12. https ://www.huffingtonpost.in/2018/12/04/there-is-no-evidence-of-a-temple-under-the-babri-masjid-asi-lied-to-the-country-say-archeologists_a_23604990/

Shrivastava, Rahul. 2018. "Review Educ Fort 'Adopted' by Dalmia Bharat Group, Congress Objects to Privatisation of Heritage." *India Today*, April 28. https://www.indiatoday.in/india/story/red-fort-adopted-by-dalmia-bharat-group-co ngress-objects-to-privatisation-of-heritage-1222339-2018-04-28

Silva, Kapila D., and Amita Sinha. 2016. *Cultural Landscapes of South Asia: Studies in Heritage Conservation and Management*. Abingdon: Routledge. https://doi. org/10.4324/9781315670041.

Singh, Rana P.B. 2016. "Urban Heritage and Planning in India: A Study of Banaras." In *Spatial Diversity and Dynamics in Resources and Urban Development: Volume II: Urban Development*, edited by Ashok K. Dutt, Allen G. Noble, Frank G. Costa, Rajiv R. Thakur, and S.K. Thakur, 423–449. Dordrecht: Springer Science + Business. https://doi.org/10.1007/978-94-017-9786-3.

Singh, Rana P.B., and Pravin S. Rana. 2017. "Varanasi: Heritage Zones and Its Designation in UNESCO' s World Heritage Properties." *Kashi Journal of Social Sciences* 7(December): 201–218.

Sinha, Amita. 2016. "The Public Realm of Heritage Sites in India: Sustainable Approaches Towards Planning and Management." In *Cultural Landscapes of South Asia: Studies in Heritage Conservation and Management*, edited by Kapila D. Silva and Amita Sinha. Abingdon: Routledge, 227–240.

Srivastava, Anupam. 2019. "Lucknow Metro Set to Cross Another Milestone." *Hindustan Times*, March 7. https://www.hindustantimes.com/lucknow/luckno w-metro-set-to-cross-another-milestone/story-X1Lyw8ljN45Mf6EMY4GO VJ.html

Srivathsan, A. 2019. "Varanasi, by Design : Vishwanath Dham and the Politics of Change." *The Hindu*, March 23. https://www.thehindu.com/society/varanasi-by-design-vishwanath-dham-and-the-politics-of-change/article26607193.ece

Teja, Charanjit Singh. 2019a. "Renovated Monuments Continue to Gather Dust." *The Tribune*, March 10. https://www.tribuneindia.com/news/amritsar/renovat ed-monuments-continue-to-gather-dust/740751.html

Teja, Charanjit Singh. 2019b. "Ram Bagh Museum Opened." *The Tribune*, May 5. https://www.tribuneindia.com/news/amritsar/ram-bagh-museum-opened/7688 79.html

Tejonmayam, U. 2017. "Metro Rail Tunnelling Has Damaged 300-plus Buildings Since Mid-2012." *Times of India*, December 1. https://timesofindia.indiatime s.com/city/chennai/metro-rail-tunnelling-has-damaged300-plus-buildings-sinc e-mid-2012/articleshow/61886194.cms

Thakur, Nalini M. 1990. "A Holistic Framework for Architectural Conservation Education." *Third World Planning Review* 12(4). http://liverpool.metapress.com /content/7696403438812g18/

Thakur, Nalini M. 2007. "Hampi World Heritage Site: Monuments, Site or Cultural Landscape?" *Landscape* 16: 31–37.

Thakur, Nalini M. 2011. "Indian Cultural Landscapes: Religious Pluralism, Tolerance and Ground Reality." *Journal of SPA: New Dimensions in Research of Environments for Living 'The Sacred'* 3. https://architexturez.net/doc/az-cf -21175

TNN. 2015. "Another Excavation Leaves Charminar Precinct Shaken." *Times of India*, June 26. http://timesofindia.indiatimes.com/city/hyderabad/Another-ex cavation-leaves-Charminar-precinct-shaken/articleshow/47823174.cms

TNN. 2018a. "Historic Porch Razed for Metro, Activists Vexed." *The Times of India*, September 21. https://timesofindia.indiatimes.com/city/lucknow/histor ic-porch-razed-for-metro-activists-vexed/articleshow/65898175.cms

TNN. 2018b. "Why All Renamings Are Not the Same." *The Times of India*, November 11. https://m.timesofindia.com/home/sunday-times/why-all-renamin gs-are-not-the-same/amp_articleshow/66571369.cms

TNN. 2018c. "Gujarat High Court Notice to ASI, Railways over Construction Near Brick Minarets." *The Times of India*, December 19. https://timesofindia.indi atimes.com/city/ahmedabad/hc-notice-to-asi-railways-over-construction-near- brick-minarets/articleshow/67168752.cms?fbclid=IwAR3WFtkB9qxzJcSy0P5p4 fpqwdopITRhhAHm4RVU4ehsrYfeSklq7yuRq-8.

TNN. 2019. "Plea Against 'Botched' Restoration at CSMT." *The Times of India*, February 26. https://timesofindia.indiatimes.com/city/mumbai/plea-against -botched-restoration-at-csmt/articleshow/68160887.cms?utm_campaign=and app&utm_medium=referral&utm_source=native_share_tray

Turaga, Vasanta Sobha. 2019. "Construction at Hyd's Historic Golconda Fort: ASI Making a Mockery of Its Own Code." *The News Minute*, March 10. https://ww w.thenewsminute.com/article/construction-hyd-s-historic-golconda-fort-asi-m aking-mockery-its-own-code-98064

Verma, Richi. 2015. "Heritage Building Falls to Delhi Metro's Tunnel Work." *Times of India*, May 21. http://timesofindia.indiatimes.com/city/delhi/Heritage- building-falls-to-Delhi-Metros-tunnel-work/articleshow/47364512.cms

Index

Note: page references in *italics* indicate figures; **bold** indicates tables.

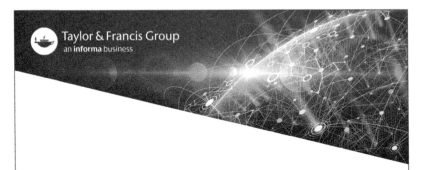

Printed and bound by CPI Group (UK) Ltd, Croydon, CR0 4YY

24/10/2024

01778306-0007